Steamboat Modernity

Constantin
Ardeleanu

Steamboat Modernity

Travel, Transport, and Social Transformation on the Lower Danube, 1830–1860

English translation by James Christian Brown

CEU PRESS

Central European University Press
Budapest–Vienna–New York

Published in 2024 by
CENTRAL EUROPEAN UNIVERSITY PRESS

Nádor utca 9, H-1051 Budapest, Hungary
Tel: +36-1-327-3138 or 327-3000
E-mail: ceupress@press.ceu.edu
Website: www.ceupress.com

First published as *O croazieră de la Viena la Constantinopol. Călători, spații, imagini, 1830–1860*, © Editura Humanitas, 2021.
Translated by James Christian Brown

On the cover: *Das k.k. privilegirte Dampfboot Maria Anna* (c. 1837) – public domain

ISBN 978-963-386-753-2 (hardback)
ISBN 978-963-386-754-9 (ebook)

LIBRARY OF CONGRESS CATALOGING-IN-PUBLICATION DATA

Names: Ardeleanu, Constantin, author. | Brown, James Christian, 1962-
translator.
Title: Steamboat modernity : travel, transport, and social transformation
on the Lower Danube, 1830-1860 / Constantin Ardeleanu; translated by
James Christian Brown.
Other titles: Croazieră de la Viena la Constantinopol. English | Travel,
transport, and social transformation on the Lower Danube, 1830-1860
Description: Budapest ; New York : Central European University Press, 2024.
| "First published as O croazieră de la Viena la Constantinopol.
Călători, spații, imagini, 1830-1860, Editura Humanitas, 2021.
Translated by James Christian Brown."--CIP title page verso. | Includes
bibliographical references and index.
Identifiers: LCCN 2024008297 (print) | LCCN 2024008298 (ebook) | ISBN
9789633867532 (hardback) | ISBN 9789633867549 (adobe pdf)
Subjects: LCSH: River steamers--Lower Danube River--History--19th century.
| Lower Danube River--History--19th century. | Lower Danube River
Valley--History--19th century. | Danube River Valley--Description and
travel. | Danube River Valley--Social life and customs. |
Travelers--Romania--History--19th century. | Romanians--Travel. | BISAC:
HISTORY / Modern / 19th Century | TRAVEL / Europe / Eastern
Classification: LCC VM461 .A73 2024 (print) | LCC VM461 (ebook) | DDC
387.2/43--dc23/eng/20240220
LC record available at https://lccn.loc.gov/2024008297
LC ebook record available at https://lccn.loc.gov/2024008298

Contents

Illustrations

Tables

Acknowledgments

This book comes out of a longstanding passion for travel and for the Danube. I began to take an interest in the theme more than two decades ago, but it was in the spring of 2020, in the unsettled period at the start of the pandemic, that the present study started to take shape. It was then that I had the time to look more closely at material gathered from a wide variety of sources and also, in a world that seemed completely at a standstill, to better appreciate the relevance of mobility. The result is this book, first published in Romanian in 2021 as *O croazieră de la Viena la Constantinopol. Călători, spații, imagini, 1830–1860* by Editura Humanitas in Bucharest.

The book is the result of my own academic voyage, for which I owe a debt of gratitude to a number of individuals and institutions.

Back in the early 2000s, my doctoral supervisor, Paul Cernovodeanu, encouraged me to turn my attention to the foreign travelers who took the Danube route. I followed his advice and prepared Romanian translations of some of their descriptions of the Romanian lands for inclusion in the volumes of *Călători străini despre Țările Române în secolul al XIX-lea* ("Foreign Travelers about the Romanian Lands in the 19th Century"). The series, published by the Romanian Academy's "Nicolae Iorga" Institute of History, has been supervised in recent years by Daniela Bușă, to whom I am grateful for an excellent collaboration. My interest in the subject also resulted in a number of courses on the Danube and its ports, which I taught until 2022 to students of the Faculty of History, Philosophy, and Theology of the "Lower Danube" University of Galați. I would like to thank both the management of the Faculty and, above all, the students, who I hope enjoyed an educational voyage through the history of the Danube region.

In recent years, I have presented various parts of my research at events organized by New Europe College (NEC) in Bucharest. Valentina Sandu-Dediu and Anca Oroveanu, the rector and academic coordinator of NEC, respectively, have been by my side in these academic excursions, as have Marina Hasnaș and the present executive director, Lelia Ciobotariu, who encouraged and supported, both personally and institutionally, the publication both of the Romanian edition and of this English translation. Over the past few years, NEC has been for me more than an institution of academic excellence: it has been a welcoming home, and I am grateful to all those who make the college such a special place in the Romanian academic world.

Wissenschaftskolleg zu Berlin was a generous host in the spring of 2022. It was through this institute's excellent library that I was able to complete my bibliography with materials I had had no access previously. My thanks to the Wiko management and in the first place to its rector, Barbara Stollberg-Rilinger.

I would also like to express my gratitude to my colleagues at the Institute of South-East European Studies of the Romanian Academy, where I have been based since 2022. My thanks to Daniel Cain and Andrei Timotin for the warm welcome with which they eased my integration into the new team.

Warm thanks are also due to CEU Press, and in particular to Jen McCall, who facilitated the whole publishing process. The evaluators appointed by the press made extremely useful observations, which have enriched this edition. I am grateful too to the management of Editura Humanitas, who agreed unreservedly to the publication of an English edition of the book.

I have kept till the end the special gratitude I owe to the translator, James Christian Brown. He has brought to the present edition not only much painstaking attention to detail but also his own research experience regarding travel literature. Thanks to him, the present volume offers the English-speaking public much of the charm of the nineteenth-century travel narratives.

Last but not least, I would like to thank my family, who waited patiently for me as I scoured libraries and archives and wandered, physically or in my mind, up and down the Danube.

November 2023

"The Character of Our Age"

"Just Like a Still for Making Raki"

Modernity could take the most diverse forms, the Wallachian boyar Dinicu Golescu (1774–1830) observed almost two centuries ago, in the course of one of his journeys through Europe (1824–1826). The towns of Transylvania, Hungary, and Austria never ceased to provide him with occasions for wonder, with the result that his record of his travels is not just a captivating description of the places he visited but also a study of the most varied human emotions. At Trieste, the great maritime entrepôt of the Habsburg Empire, the boyar's admiration was at its peak. As he wandered the streets of the famous Adriatic port, he found himself surrounded by a reality so remarkable that "it is no good hearing it described, it has to be seen." "The beauty of the streets, the lines of the houses, the edges of the sea full of boats, the hills with vegetable gardens and vineyards"[1] were just a few of the most picturesque images of this town animated by a huge entrepreneurial energy.

The port was the heart of Trieste, frequented annually by hundreds of vessels bringing people and goods from far-flung places. The ships known as *curiere* departed daily for Venice and offered a weekly connection with the great ports of the Mediterranean. There was also a *vapor* (steamboat) that circulated twice weekly between Trieste and Venice. This state-of-the-art vessel seemed

---

[1] Constantin Golescu, *Însemnare călătoriei mele în anii 1824, 1825 și 1826*, ed. Petre V. Haneș (București, 1915), 104–105. The English translation is taken from Alex Drace-Francis's fragment of *"Dinicu Golescu, Learning from Enlightened Europe (1826),"* in Wendy Bracewell (ed.), *Orientations: An Anthology of East European Travel Writing: ca. 1550–2000* (Budapest, 2009), 101–103. For the relevance of Golescu's account, see Drace-Francis, *The Traditions of Invention: Romanian Ethnic and Social Stereotypes in Historical Context* (Leiden, 2013), 135–158.

to be a great success. "Such a boat always has, in addition to the goods and the crew, from forty to sixty passengers, and on occasion up to a hundred; so you can imagine what an uninterrupted flow of masses of people traveling!"[2]

Curiosity drove Golescu to embark on the *vapor* and take a trip to the city of lagoons. "The *vapor*," he explains in a detailed note,

> is a boat which goes on the sea by means of a contraption of fire, which is in the cabin of the boat, while on the outside only an iron chimney, about four *stânjeni*[3] long, out of which the smoke comes, and two big iron wheels, just like mill wheels, which go in the water, one on one side, outside the boat, and the other on the other side, on an iron axle, about six hands[4] higher than the water surface; a part of the wheels goes into the water and three parts remain outside.

The movement of the paddles propelled the steamboat with such great force "that one's body feels the boat smashing against the sea, and the wheels leave behind two long tails of foam." The system was not completely foreign to his compatriots, the curious boyar further remarks in his technical explanations: the stove constructed in "the cabin of the vessel" was provided with several iron chimneys, one of them directed

> toward the mechanism they have, through which heat and steam emerge just like a still for making raki, where a fire burns underneath and a drop of steam gathers at the lid. Thus the steam from that pipe moves the first wheel, where there are perhaps three times as many [missing word] as on a clock, and the second wheel, with its teeth and the teeth on the iron axle, turns the axle and both the wheels.

Also using "masts with canvas," when the wind was favorable, the steamboat flew across the waters of the Adriatic Sea.[5]

Speed and comfort were two of the acclaimed advantages of the steamboats that had recently begun to appear on the seas and rivers of the Western

---

2   Golescu, *Însemnare*, 105–106; Bracewell, *Orientations*, 103–104.
3   About 8 m.
4   About 1.5 m.
5   Golescu, *Însemnare*, 105–107, n. 1; Bracewell, *Orientations*, 103–104, n. 2.

world. Golescu traveled to Venice by *vapor*, but he chose to return to Trieste on a sailing ship. He was soon to "curse" his wish to compare the propulsion methods of the two vessels. The journey back by sail took forty hours, "and for the whole that time, I had no food, no sleep, and did nothing but throw up and cry like a baby."[6] The benefits of modern technology were appreciated by the passenger's stomach as much as by his mind.

The "Machine Age": "Wonders of the Time" or "Work of the Devil"

In the years that followed, more and more of Golescu's compatriots in the Danubian Principalities of Wallachia and Moldavia would travel through Europe, becoming familiar not only with, for example, the good organization of administrative systems and education in various Western countries but also with the advantages brought about by the transport revolution. Twenty years younger than Golescu, Petrache Poenaru (1799–1875) was studying engineering in Paris when, in 1831, he visited Britain. In London, Poenaru admired "the imposing bridges [over the Thames], which, although they are built on a terrifying scale, nevertheless seem so graceful," and the beautiful streets "so clean that they seem to give more health to the air you breathe." However, one of the most memorable moments of his stay in Britain was a journey by train between Liverpool and Manchester, which he made in October 1831, just one year after the inauguration of the first railway that relied exclusively on mechanical propulsion. The journey was made, Poenaru noted, "in a totally new manner, which is one of the wonders of the time [...]. Twenty wagons, connected to one another and carrying 240 persons, are pulled all at the same time by a single steam engine, and the train advances so fast that the best racehorse could not follow it at a forced gallop."[7]

As these two Wallachian patriots had experienced for themselves, humanity was entering a new phase of historical development, and steam was one of the motors of change. Industrial workshops were accelerating the production of goods, and the new means of transport propelled by steam engines—steamboats and trains—were making travel more comfortable, faster, and

---

6   Ibid.
7   *Scrisori vechi de studenți (1822–1889)*, ed. Nicolae Iorga (București, 1934), x–xi.

safer.[8] For the author of an article in the newspaper *Curierul românesc* ("The Romanian Courier," 1829–1830), the conclusion regarding "the character of our age" was clear: a "mechanical age" had begun, in which the *piroscaf,* as the steamer was termed under French or Italian influence, was replacing the sailing vessel and the steam carriage was taking the place of the horse.[9] Three great discoveries, another Wallachian author remarked in *Albina românească* ("The Romanian Bee," 1837), had influenced "the destiny of people more than the revolutions of the earth and the collapse of empires," namely "the printing press, the compass, and steam." Each of these had opened up new prospects for the mental and physical progress of humanity, with major consequences also for the "social system" of the world.[10] For the authors of these articles, the relation between material progress and civilization was abundantly clear, in a way almost suggestive of the concepts later developed by Fernand Braudel, Norbert Elias, and the sociotechnical systems theorists.[11]

The steam age had come, and, finding themselves at a favorable historical juncture, the Wallachians and Moldavians were quick to come to terms with the new technical discoveries and, through them, with modern spaces and times. Encountered first of all abroad, through the accounts of such witnesses as Golescu and Poenaru, the transport revolution arrived in Wallachia and Moldavia (Figure 1) with the first Austrian steamer that began to stop at the Principalities' Danube ports in April 1834. A note published in a number of Western newspapers two years later, the initial source of which remains unknown, maintains that, on their first sight of a steamboat, the local inhabitants fled as fast as their legs could carry them, convinced that the contraption was "the work of the devil."[12] The anecdote brings to mind other equally memorable encounters between populations with a rudimentary level of technology and modern vehicles. One such was recounted by the British engineer Henry C. Barkley (1837–1903). The completion of the

8   For the general context, see Philip Bagwell, *The Transport Revolution, 1770–1985* (London, 1988).
9   *Bibliografia analitică a periodicelor românești* (hereafter BAPR), vol. 1, *1790–1850,* part 1, ed. Ioan Lupu et al. (București, 1966), 3 ("Haracterul epohi noastre," *Curierul Românesc* 1, 1829–1830).
10 Ibid., part 3 (București, 1967), 879 ("Vasul de vapor," *Albina românească* 8, 1837).
11 Fernand Braudel, *Civilization and Capitalism, 15th–18th Century,* vol. 1, *The Structures of Everyday Life,* trans. Siân Reynold (Berkeley, 1992); Wiebe Bijker and John Law (eds.), *Shaping Technology/Building Society: Studies in Sociotechnical Change* (Cambridge, MA, 1992); Norbert Elias, "Technization and Civilization," *Theory, Culture and Society* 12, no. 3 (1995), 7–42.
12 Anon., "Navigation of the Danube," *Army and Navy Chronicles* 2–3 (1836), 104.

railway and the circulation of trains between Cernavodă[13] and Constanța in 1860 left the Turks of Dobrogea relatively indifferent. One local explained to Barkley the reason for their lack of interest in the locomotive: the people did not want to have anything to do with the unclean one, and they believed that "a strong young devil" was shut up "in that great fire-box on wheels, where you induce him to turn a crank connected with the wheels, and pay him for doing so by giving him cold water to allay his tortures."[14] A second anecdote concerns the arrival of the first train in the Transylvanian town of Deva in September 1868, an event that caused a sensation in the town and its surroundings. "The peasants and townsfolk saw in the fire-cart a veritable devilish miracle, while the gentry were astonished and could not understand how it ran without being pulled or pushed by something."[15] The sensory impact of the technology propelled by the force of steam, with its lights, temperature, smell, and, above all, the noises[16] produced by the powerful engines, obviously favored such diabolical associations.

The scholar monk Chiriac (1793–?) of Secu Monastery in Moldavia was one of the most curious travelers in the period examined in this book. In the following chapters, I shall refer to some of his journeys through Europe. In November–December 1850, for example, returning to his homeland after traveling "through Russia, Lehia [Poland], Nemția [Germany], Hungary, and Transylvania," Chiriac went between Warsaw, Krakow, Vienna, and Budapest by "steam" (*vapor*) on land and on water.[17] His fascination with science and technology made him take a note of all sorts of information of interest to his readers, "for each to know how many roads a kingdom has and how much it spends on making them, but also how much benefit this brings." In the course of his journey to Budapest, the train passed through a tunnel.

---

13 When not in quotations, place names will be spelled in their current form. Exceptions will be made for places where the English equivalent is commonly used, such as Bucharest for București.

14 Henry C. Barkley, *Between the Danube and the Black Sea; or Five Years in Bulgaria* (London, 1876), 260–261.

15 Constantin Botez, Dem. Urma, and Ioan Saizu, *Epopeea feroviară românească* (București, 1977), 55.

16 Karin Bijsterveld, "The Diabolical Symphony of the Mechanical Age: Technology and Symbolism of Sound in European and North American Noise Abatement Campaigns, 1900–40," *Social Studies of Science* 31, no. 1 (2001), 37–70.

17 His use of the Romanian word *vapor* (nowadays used only of ships) to refer to trains may seem strange to Romanian speakers today. However, setting aside the different physical infrastructure on which these vehicles circulated—on the "road of iron" or on the "road of water"—the technology by which they were propelled was the same.

Figure 1 *Map of the Course of the Danube from Ulm to Its Mouth in the Black Sea* (1837).

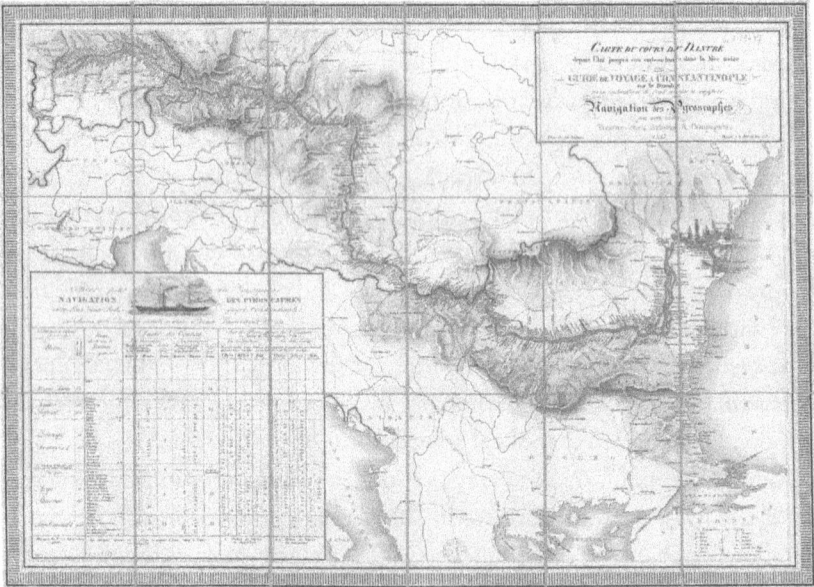

And very much I wondered that these emperors do not stop making roads even through mountains, that is, under the ground, for I had heard before, but I did not believe, but now I have seen and I believe. For mountains thus cut I had seen before, just as on this road there were many cut in places, even 6–7 *stânjeni* [*c.*12–14 meters], but mountains pierced through I had never seen until now.[18]

The majority of contemporaries must have agreed with such remarks on the contribution of modern means of transport to not only the physical but also the political, economic, and social transformation of the world. Like the iron roads, remarks the author of an article in the newspaper *Romania* (1838), "floating by steam [...] can be seen to be destined to change all the relations of countries and of nations on the face of the whole earth."[19]

18 G. Giuglea, "Călătoriile călugărului Chiriac de la Mănăstirea Secul," *Biserica Ortodoxă Română* 54, nos. 11–12 (1936), 702, 705–706.
19 BAPR, vol. 1, part 3, 886 ("Plutirea cu vapor," *Romania* 1, 1838).

The force of steam, another author noted two years later in the Iași paper *Icoana Lumei* ("The Image of the World"), announced entry into a new age "for industry and the advance of civilization." The introduction of steam navigation on the Danube contributed not only to the development of Moldavia's commerce "but also [for] civilization there has opened a facility that will produce the most fortunate results."[20] Traveling by train in the Habsburg Empire in 1844, Moldavian intellectual Mihail Kogălniceanu (1817–1891), who would later serve as prime minister of modern Romania in the 1860s, comments on the fact that "without wishing" Austria was "civilizing and bringing liberty, for, by way of the iron roads, new ideas can be spread."[21]

Such correlations between the force of steam and civilization were commonplace in the period. Belief in the modernizing power of steam had become a sort of ideology for those who were convinced that the new technologies would change the world. Numerous authors in Wallachia and Moldavia wrote about the relation between technology and civilization, including Simion Marcovici (1802–1877) as well as Kogălniceanu.[22] Almost all praised the advantages of the modern means of transport as factors of progress and bemoaned the poor state of public roads in their native lands.

Railways were the most obvious carriers of modernity. With them, and by them, came movement, commerce, civilization. Economist and statesman Ion Ghica (1816–1897), another prime minister of modern Romania, noted that they contributed to the moral unification of the European nations and to the taking root of the idea of common identity as peoples. Through them "is facilitated the development of commerce, of civilization, of economic prosperity, and the cementing of tighter bonds between nations." Railways were not only transport infrastructure but also had "a mission even greater, they exert a moral and political action upon those societies through which they pass: they have become the most powerful agent of civilization and freedom, bringing and implanting the idea and the spirit of one people to the others." Kogălniceanu, too, spoke of material prosperity and morality, as "the law of the iron road is to overturn and to level any impediments of

---

20 Ibid. ("Vasul cu vapor," *Icoana lumei* 1, 1840).
21 Mihail Kogălniceanu, *Opere*, vol. 1, *Beletristica, studii literare, culturale și sociale*, ed. Dan Simonescu (București, 1974), 488.
22 For more details, see Ștefan Pascu, Radu Pantazi, and Teofil Gridan, *Istoria gîndirii și creației științifice și tehnice românești* (București, 1982).

whatever sort they may be."[23] Poet Alecu Donici (1806–1865) offered, in the service of the struggle for the unification of the Principalities, an interesting fable entitled "The Steam [Locomotive] and the Horse" (*Vaporul și calul*). In response to a horse looking in amazement at this dragon that runs whistling "on the road with iron rails," the locomotive reveals the secret of its predominance: "Power through union."[24]

Such ideas about the relation between the power of steam and civilization were common among the elite in the Danubian Principalities, as Vasile Alecsandri (1821–1890) tells us with subtle irony through the voice of the most famous of the heroines of his comedies, Chirița Bârzoi or "Coana Chirița." Like Golescu or Poenaru, Chirița makes no secret of her enthusiasm for that "wonderful thing" the steamboat, "fire on water" that flies somewhat as in the time of the *Halima*.[25] It is true that the locomotive seemed an even greater wonder, which had revolutionized transport so much that Chirița had heard that "in England the iron road goes so fast that it arrives before it leaves." "Bravo to the countries that have iron roads!" Alecsandri's heroine also exclaims, with a conclusion about the relation between transport infrastructure and civilization: "They have wings to fly fast on the road of progress, but as for the rest, as in our country, for example, they are handless and legless, the poor things!"[26]

Received variously with enthusiasm or curiosity by the mobile strata of society, and with indifference or fear by the simple folk, the mechanical wonders or diabolical technologies that moved the world quickly transformed the lives of the inhabitants of the Principalities of Moldavia and Wallachia too. As was well understood by all those who had traveled on steamboats and trains shortly after their introduction in the Western world, the revolution in transport meant more than being able to move at a speed higher than that of a racehorse. The new vehicles were transforming everything around them: space and time, material civilization and social relations, the business environment and the natural environment. The world was changing before people's eyes.

---

23 Qtd. in Botez et al., *Epopeea*, 23–24.

24 *Acte și documente relative la istoria renascerei României*, vol. 3, ed. Ghenadie Petrescu, Dimitrie A. Sturdza, and Dimitrie C. Sturdza (Bucuresci, 1889), 615.

25 "Halima" was the name under which the collection of Arabic tales *The Thousand and One Nights* had become popular in the nineteenth century. By extension, the term had the sense of something wonderful, but also of a story full of adventures.

26 Vasile Alecsandri, *Opere complete*, vol. 2, *Teatru I* (Bucuresci, 1903), 44–46.

The fact that the transport revolution reached Wallachia and Moldavia so rapidly resulted from the location of the Principalities at a crossroads of empires and on a navigable river, a natural highway that lacked only the most efficient technology to become a fast, safe, and comfortable connection between West and East. The steamboat filled this technological void just at the time when a commercial revolution was showing that the territory of the Principalities had many more resources to offer.

THE AGE OF THE STEAMBOAT

The 1829 Treaty of Adrianople (now Edirne, Turkey) between Russia and the Ottoman Empire marked the beginning of a commercial revolution in the history of the Lower Danube region. New business opportunities, generated by the abolition of the commercial restrictions previously imposed by the Sublime Porte, now opened for Moldavia and Wallachia. The external trade of both Principalities had hitherto been dominated by the export of certain "strategic" products (mainly grain and cattle), often at imposed prices, to stock the Ottoman market. Other goods that were found in abundance in the empire could be exported without restriction, but the proportion of these was relatively small in the total volume of the Principalities' commerce. A gradual change had begun in the 1770s, with the development of international trade in the Black Sea basin. Later, after the annexation of Bessarabia by the Russian Empire in 1812, the Danube ports of Izmail and Reni became the centers of an increasingly significant trade in grain, hides, suet, wax, and so on.

For contemporaries, the Treaty of Adrianople marked the end of the Ottoman "monopoly" and the beginning of a period of commercial liberty for the Principalities, based on the principle of free trade. The role of imperial Russia in the genesis of capitalism on the Lower Danube is, it must be recognized, one of the ironies of Romanian modernity. A series of financial and customs measures adopted by the authorities in Bucharest and Iași, including granting the status of free ports to Brăila and Galați (1836–1837), transformed the Principalities into an attractive economic space for Mediterranean and Western merchants interested in the commercial opportunities offered by a region that was rapidly becoming connected to the global capitalist system.

The exponential growth in grain exports called for a transport infrastructure to match, and the Danube was the route on which the prosperity of the Principalities depended. The French politician and writer Saint-Marc Girardin (1801–1873), who traveled on the Danube on board the Austrian steamer in 1836, captured in a synthetic formula the interdependence of trade, navigation, the Danube ports, and the prosperity of the Principalities: "Free trade needs the navigation of the Danube, for the river links the Principalities upstream to Central Europe, and downstream to the Black Sea and the Mediterranean. Two towns, Brăila in Wallachia and Galați in Moldavia, personify, so to speak, the interests and hopes of the Principalities with regard to the Danube."[27] One of the metaphors commonly used with reference to the integration of rivers and other navigable routes within modern states is that of circulation. The theory of the circulation of the blood, formulated by the English doctor William Harvey, was taken up by Thomas Hobbes, who made much use of it with reference to matters of state.[28] The term "artery" was also adopted in the field of transport to designate those communication routes that were vital for the functioning of the state organism. In the Romanian space, anatomical comparisons also went in a different direction, with the ports of Brăila and Galați being seen as the lungs of the Principalities, thus making the Danube their respiratory tract, by way of which the organism received the oxygen necessary for survival.

I have examined elsewhere[29] the development of the international commerce of the two states, the growth of their export capacity and of the interest of foreign merchants in Danubian products. This is not the place to enter into details about such aspects as the structure, value, or orientation of foreign trade through the Danube ports, the qualitative and quantitative development of navigation on the Danube, how circulation at the mouth of the river and on the Sulina branch became a diplomatic issue, or the political, legal, and technical solutions to eliminate the obstacles that prevented navigation.

---

27 Saint-Marc Girardin, *Souvenirs de voyages et d'études*, vol. 1 (Bruxelles, 1852), 238.

28 Edit Király, *"Die Donau ist die Form": Strom-Diskurse in Texten und Bildern des 19. Jahrhunderts* (Wien, 2017), 88–91.

29 See the monographs: Constantin Ardeleanu, *Evoluția intereselor economice și politice britanice la gurile Dunării (1829–1914)* (Brăila, 2008); Ardeleanu, *Gurile Dunării—o problemă europeană. Comerț și navigație la Dunărea de Jos în surse contemporane (1829–1853)* (Brăila, 2012); Ardeleanu, *International Trade and Diplomacy at the Lower Danube: The Sulina Question and the Economic Premises of the Crimean War (1829–1853)* (Brăila, 2014).

What is important for the present study is the fact that the beginning of the circulation of steam vessels on the Danube was concomitant with the massive penetration of the international cereal trade by Danubian grain. The relation between these two processes was a close one, and, as I shall show below, commercial factors carried considerable weight in the plans of Austrian decision-makers to extend steam navigation to the mouths of the Danube and into the Black Sea. However, I should make it clear that until the Crimean War, steam shipping played a significant role, at least on the Wallachian and Moldavian stretch of the Danube, in revolutionizing the transport of passengers, parcels, currency, and information, rather than that of goods. The result that should be borne in mind is that the ports of the Lower Danube became part of a complex network of inter-imperial transport, with Austrian steamboats ensuring a rapid connection between East and West. The success of the grain trade was also due in no small measure to this connection, which ensured the mobility of merchants and information to and from the great consumer markets of industrialized Europe.

If, in the last third of the nineteenth century, Romania laid more than 3,000 km of railway track, the transport infrastructure that supported the economic development of the Old Kingdom[30] in the "age of the train," the middle third of the same century (broadly speaking the years 1830–1860), may be considered the "age of the steamboat," the means of transport that connected the Danubian Principalities to the rest of the world. I shall offer sufficient arguments in the book to support this view. For now, I limit myself to mentioning that on the economic and political level, from the point of view of the circulation of people, goods, and ideas, through the number of foreigners who arrived in the Principalities and of Moldavians and Wallachians who traveled either West or East, steamboats—initially Austrian, followed later by those flying the flags of imperial Russia, France, or Britain—were crucial instruments for the modernization of Romanian

---

30 The Romanian "Old Kingdom" refers to the territory forming modern Romania from its independence (1878) and recognition as a kingdom (1881) until its territorial expansion after the First World War. It included the two Principalities, Wallachia and Moldavia, which united in 1859, as well as the province of Dobrogea. For simplicity, I shall use "Romanian" to refer to realities specific to the two Principalities and their inhabitants for the period that preceded the making of modern Romania. For a modern approach to the modernization of the Principalities and the making of a unified Moldo-Wallachian citizenship, see Constantin Iordachi, *Liberalism, Constitutional Nationalism, and Minorities: The Making of Romanian Citizenship, c. 1750–1918* (Leiden, 2019), 31–191.

society. Just as cereal exports provided the financial resources on which Romanian modernity was built and the Western model provided the pattern on which modern institutions and practices were based, so the steamboat was the principal vehicle by which people, information, and ideas circulated toward the two Principalities in the phase in which they were strengthening their connections with the rest of the world, after the sporadic contacts of previous centuries.

Trade Route and State

A preoccupation common to the academic work of three of the greatest Romanian historians of the twentieth century—Nicolae Iorga, Gheorghe I. Brătianu, and Şerban Papacostea—concerns the role of the great intercontinental trade routes in the genesis of the medieval states of Wallachia and Moldavia. To paraphrase the title of one of Iorga's books, we may recognize that the trade routes were the creators of the medieval states of Wallachia and Moldavia. As political entities, Iorga considered, the two states contributed to the security of international commercial exchanges along vital sectors of the routes connecting the center and north of Europe with the mouths of the Danube and the Black Sea. Brătianu took up the theory, bringing new arguments on the basis of which he concluded that, in their early days, the medieval states established themselves as guardians of these great trade routes and that "here the route was able to create the state." Papacostea examined the claims of his predecessors critically and concluded that "the route did not 'create' the state," even if the commercial arteries and disputes over their control were "a decisive component of the final stage of the formation of the feudal state, both in the case of Wallachia and in that of Moldavia." It is, however, true, continues Papacostea, that the trade routes precipitated the territorial formation of the two states, favored their urban development, and fixed the principal directions of their foreign and commercial policies for a century and a half, until the establishment of Ottoman control over the region of the Lower Danube and the Black Sea in the fifteenth and sixteenth centuries.[31]

My aim in the present volume is to use the same theoretical perspective to examine the function of another route in the formation of modern

---

31   Qtd. from Şerban Papacostea, "Drum şi stat," *Studii şi materiale de istorie medie* 10 (1983), 9–55.

Romania in the period between 1830 and 1860. One of the main premises of the book is that through the introduction of steam navigation on
the Danube, the great river became an important artery of intercontinental transport, making the Principalities of Wallachia and Moldavia key
territories in ensuring a safe, fast, and comfortable connection between
East and West. That cliché of Romanian historiography, the placing of the
Principalities "between Orient and Occident," to quote the title of a popular book by Neagu Djuvara,[32] is very appropriate, at least from the point
of view of the construction of pan-European transport infrastructure in
the nineteenth century.

Thus, the natural "highway" of the Danube, served by the vessels of two
Austrian steamboat companies, permitted an acceleration in the circulation
of people and goods along a route that is indeed fascinating in its own right.
Increased mobility on the river, with thousands of passengers passing along
the Lower Danube on their way to Vienna or to Istanbul, contributed not
only to an exponential growth in the interest shown by foreigners in the economic, political, and social situation of the two Danubian Principalities but
also to a multiplication of the transport options available to the local population to explore destinations in Europe and beyond. In other words, the
navigable route of the Danube facilitated the meeting of Moldavians and
Wallachians with the world and the connection—economic and political,
but also mental—of the Principalities with European civilization. As part
of a complex international transport infrastructure, the Lower Danube and
the Austrian steamboats that navigated it favored the formation of modern
Romania (Figure 2), just as, according to Iorga, international trade routes
had contributed to the genesis of the medieval states.

My aim in this book is to turn the spotlight on this aspect that has
been little researched in Romanian historiography: the routes by which
the Romanian lands became connected to Europe. I propose to write the
history of a waterway, with some of the aspects arising from the circulation of travelers and goods along it: matters of infrastructure and environment, means of transport and stopping places, technological progress and
forms of mobile socialization. Along the navigable waterway of the Lower

---

32 Neagu Djuvara, *Le pays roumain entre Orient et Occident: les principautés danubiennes au début du XIXe siècle* (Paris, 1989).

Figure 2 J. E. Cadiou, *Map of the United Principalities (Romania)* (1864).

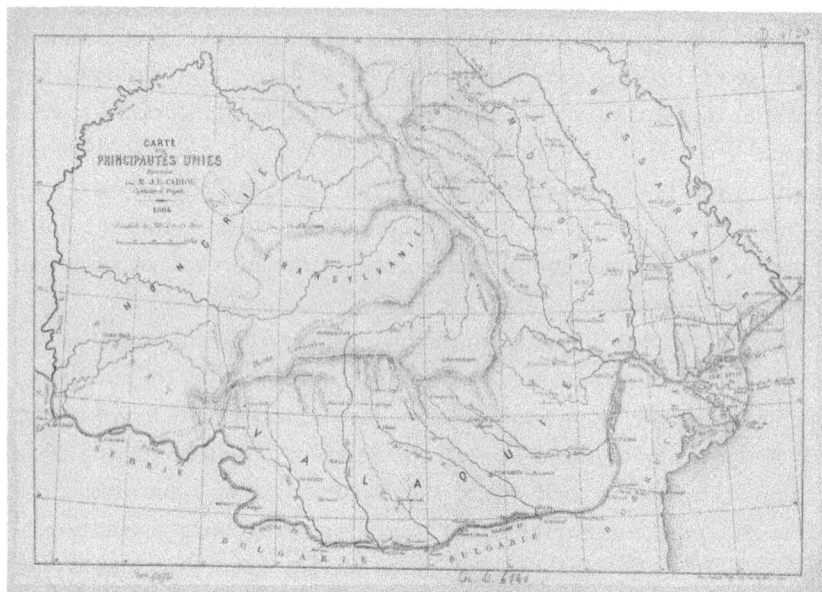

Danube, the circulation of steamboats favored urban development and contributed directly to the modernization of some of the institutions of the Principalities (the public transport, quarantine, fiscal, and customs systems). As a sociotechnical system (made up of the physical subsystem of transport and the socio-human subsystem of mobility, interactions, and other human activities),[33] the Danube route integrated the Romanian space in a dense network of European roads and waterways. Not only was the Danube the backbone of regional transport infrastructure throughout the period covered in this book, but it also determined the later structure of the national transport network. As John J. Jensen and Gerhard Rosegger have shown, the construction of the first railways in the Romanian space outside the Carpathian arc was possible precisely because the Black Sea and

---

33 Paul N. Edwards, "Infrastructure and Modernity: Force, Time, and Social Organization in the History of Sociotechnical Systems," *Modernity and Technology* 1 (2003), 185–226; Bill Hillier, "The City as a Sociotechnical System: A Spatial Reformulation in the Light of the Levels Problem and the Parallel Problem," in S. M. Arisona, G. Aschwanden, J. Halatsch, and P. Wonka (eds.), *Digital Urban Modeling and Simulation: Communications in Computer and Information Science* (Berlin, 2012), 24–48.

the Danube facilitated the importation of the necessary technology.[34] This natural transport artery played a vital role in the establishment and modernization of Romania; however, its importance was gradually eroded as the development of terrestrial infrastructure, and of the great sea port of Constanța, transformed the country into a more dynamic economic organism and one better adapted to the challenges of a world that had stepped into a new stage in the transport revolution.

FIVE HETEROTOPIAS AND NUMEROUS TRAVELERS

This book is the fruit of two lines of research that I have pursued throughout my academic career: the history of the Danube, and travel literature. Both benefit from a rich historiography, to which I shall make frequent references. In this introduction, I shall draw attention to just three authors of recent studies directly connected to the theme of my book. Luminita Gatejel has applied the perspective of STS (science, technology, and society) studies to the complex works of hydraulic engineering at the Iron Gates, giving a detailed account of how the Lower Danube came to be integrated among the routes of international mobility.[35] In a comprehensive study with excellent theoretical grounding, Edit Király has examined the way in which, in the course of the nineteenth century, the Danube acquired its identity as a regional connector. She gives particular attention to the texts that invented the tradition made famous a century later by the writers Claudio Magris and Péter Esterházy, showing how, in the context of its gradual distancing from the German space, the Habsburg Empire metamorphosed into the "Danubian monarchy."[36] Most recently, in two studies based on archive research and travel literature, part of a doctoral thesis on the subject, the

---

34 John H. Jensen and Gerhard Rosegger, "British Railway Builders along the Lower Danube, 1856–1869," *Slavonic and East European Review* 46, no. 106 (1968), 105–128; Jensen and Rosegger, "Transferring Technology to a Peripheral Economy: The Case of the Lower Danube Transport Development, 1856–1928," *Technology and Culture* 19 (1978), 675–702; Alina Cristina Munteanu, "Travelling in Oriental Romania in the Second Half of the Nineteenth Century, According to the Writings of Western Travellers," *Studia Universitatis Petru Maior. Historia* 15, no. 1 (2015), 15–24.
35 Luminita Gatejel, "Overcoming the Iron Gates: Austrian Transport and River Regulation on the Lower Danube, 1830s–1840s," *Central European History* 49, no. 2 (2016), 162–180. See also Gatejel's most recent book, *Engineering the Lower Danube: Technology and International Cooperation in an Imperial Borderland* (Budapest, 2022).
36 Király, *"Die Donau ist die Form."*

Hungarian historian Tinku-Szathmáry Balázs has dealt with the connection between Vienna and Istanbul and the various experiences of passengers on board the Austrian steamboats.[37] Together with similar studies focusing on other parts of the world, these testify to a growing interest in mobility along the waterways and in the way in which, starting in the nineteenth century, steam navigation transformed time and space for the territories that benefitted from the regular services of steamboat companies.

Although travel literature is the principal source of the present study, I have also, where appropriate, made use of primary sources of other types (archival documents, press articles) to better evoke the context of the age. I have kept a significant element of narration and anecdote, in the hope of offering not only historical rigor but also enjoyable reading. The text is thus sprinkled with substantial quotations, which indeed come with much of the subjectivism inherent in the description of experiences as personal as travel. This subjective and relative note will, I hope, help capture something of the charm of a period when the industrialization and commodification of travel was just beginning.

The six chapters deal with a variety of aspects relevant to the history of the Danube steamboat services. Five of them refer to different spaces (the river, the steamboats, the riverside territories, the lazarettos, the Iron Gates), as they were reimagined and given new value after the introduction of steam navigation, while the sixth presents the journeys of Romanians (i.e., Wallachians and Moldavians) by Danube steamboat and by other means of transport on land. Ships are classic examples of heterotopias in Michel Foucault's influential study.[38] However, as I shall try to demonstrate in the chapters that follow, the industrialization and commodification[39] of travel in the nineteenth century created other discursive spaces too. These special regions made travel on the Danube and the Black Sea all the more memorable an experience.

The first chapter presents the context in which the Austrian steamboat service was introduced on the Danube and the organization of the transport

---

37 Balázs Tinku-Szathmáry, "Gőzhajóval a Dunán Bécsből Konstantinápolyig," *Közlekedés-és technikatörténeti Szemle* (2018), 11–38 and (2019), 9–44.

38 Michel Foucault, "Of Other Spaces," *Diacritics* 16, no. 1 (1986), 22–27.

39 With reference to the emergence of the capitalist system, "commodification" means the transformation of objects, services, or ideas into goods destined for commercialization.

network between Vienna and Istanbul. In a historical approach to the geography of transport, I describe traveling conditions on the overland routes of the Ottoman and Habsburg Empires, identifying the components of the transport infrastructure in southeastern Europe in the second quarter of the nineteenth century. Details about the "Wallachian branch" of the direct route between Istanbul and Vienna situate the Romanian lands in the network of international transport corridors. A short history of the Austrian steamboat company, not forgetting the other actors and factors that enabled its development, shifts the story to Vienna, following which, returning downstream, I present the principal nodes and connections of the network, with particular emphasis on the Lower Danube. The chapter closes with a look at the structural changes in the pan-European transport network after the 1860s, when the development of railway infrastructure in southeastern Europe led to a gradual decline in the importance of the Danube route in ensuring connections between West and East.

In the second chapter, the Austrian steamboats serve as arenas of global history, by way of which I explore the social dimension of travel. I first of all discuss the forms of sociality created by the introduction of steam navigation on the Danube. Sailing along a waterway that was at the same time both a border and a highway connecting the Austrian, Ottoman, and Russian Empires, the steamboat was, as writer Alexandru Pelimon (1822–1881) noted, itself a "little planet"[40] with an intense social life. As sociality is an intrinsic component of modern transport, the chapter highlights aspects of daily life illustrative of the early phases of the industrialization and commodification of travel, placing the accent on the social experiences that passengers had in the course of their voyages between empires and civilizations.

The third chapter is about perceptions, clichés, and the imaginary, examining the way in which the image of the two Principalities was sketched in the descriptions of travelers who came to "know" the territory to the north of the Danube at steamboat speed. As they spent a few days passing along the Danube border and gathered information by empirical observation (the Moldavians and Wallachians on board, locals at harbors, short visits to Giurgiu when the steamboat stopped, or to Galaţi during the transfer from river to sea transport), passengers formed clear opinions about the situations of Wallachia and

---

40 Alexandru Pelimon, *Impresiuni de călătorie în România*, ed. Dalila-Lucia Aramă (Bucureşti, 1984), 138.

Moldavia. Their accounts include ample references to the political, economic, and social landscape of the Principalities, and in the course of this chapter, I synthesize some of the favorite themes in their descriptions, which seem to be based on a mixture of direct observations and clichés recycled from guidebooks and travel literature the authors had read.

The steamboats of the DDSG (Donau-Dampfschiffahrts-Gesellschaft—the Danube Steam Navigation Company) accelerated mobility in southeastern Europe as a whole, but almost nowhere along the great river did they produce more revolutionary effects than in the two Principalities. In the fourth chapter, I turn to the way in which Romanian authors perceived the steamboat and the experience of steam travel both to the West and to the East. I present the preferred itineraries and the journeys of a number of Romanian travelers, the advantages and inconveniences of the Danube route in comparison to roads on land, the cost of river travel, their various experiences on board, and the role of the steamboats as "revolutionary" vehicles.

The fifth chapter is dedicated to an important obstacle in the way of mobility on the Danube, namely quarantine regulations. With outbreaks of disease all the more frequent and virulent in the region, a passenger on the Austrian steamboat would encounter a variety of approaches to the nature and transmissibility of illness. In the Danube lazarettos, which will be discussed as contact zones, passengers became familiar not only with strict procedures to combat and prevent epidemics but also with various ways of ensuring the sanitary and political cleanliness of the bodies both of passengers and of the Principalities. The different sanitary visions also resulted from scientific disputes regarding the source of the spread of diseases, with specialists (not to mention the interested public) being divided between adepts of contagionist and of miasmatic theories.

The last chapter is about nature and technology, both of them major "discoveries" of the nineteenth century. I shall insist upon the way in which the Iron Gates region (Romanian: *Clisura Dunării*)—the gorge that now separates Romania and Serbia—was imagined and transformed following the introduction of the steamboat service. The perspective of the engineers who used modern technology to facilitate the circulation of steamboats will be complemented with references both to the political deals that allowed the hydraulic works to go ahead and to the rites of passage through a symbolic "gate" linking West and East. I shall also touch on the aesthetic component of this special space, which is increasingly present in literary treatments and artistic representations, in a classic experiment in the invention of landscape at the meeting of nature and civilization.

I thus hope not only to contribute to better understanding of one of the principal routes by which the Romanian lands were connected to the world but also to unveil the multiple transformations produced by the contact of the Romanians and of the Principalities with the modern technology that circulated along a great international transport artery, in a period when this was one of the easiest, fastest, safest, and most comfortable connections between West and East.

A Cruise between Civilizations

THE BEST CHOICE?

In spite of numerous inconveniences—crowding, mosquitoes, fear of the shivers of malaria—the Austrian steamboats that circulated regularly on the Black Sea and the Danube were, at the beginning of the 1840s, the best choice for a journey between Istanbul and Vienna. This was the conclusion reached by the English doctor Octavian (John) Blewitt in the torrid days of late July 1840. Blewitt (1810–1884) was secretary of the Royal Literary Fund, an institution that supported the publication of the works of British writers, and was himself a prolific author of travel literature, which he signed with the pseudonym "Brother Peregrine." He also edited several volumes in the "Murray's Handbooks for Travellers" series, which was extremely popular in the period (see below); so his option was that of a relatively well-informed traveler.[1]

Three main routes linked the two imperial capitals: by sea via Trieste, overland through Belgrade, and by the Danube and the Black Sea. According to the 1840 edition of one of Murray's guides, the steamboats of the Östereichischer Lloyd company offered two sailings per month between Istanbul and Trieste (Figure 3). The voyage, stopping at Izmir, Syros, Piraeus, Patras, Corfu, and Ancona, took some fourteen days. From Trieste to Graz (via Ljubljana), there were three mail coaches per week, covering the distance of some 330 km in fifty hours. Between Graz and Vienna, there were five

---

1   Alexis Weedon, "Blewitt, (John) Octavian (1810–1884), Writer and Literary Administrator," in *Oxford Dictionary of National Biography* (Oxford, 2004), https://doi.org/10.1093/ref:odnb/2645 (accessed August 28, 2023).

Figure 3 A. H. Dofour, *Map of the Eastern Mediterranean and Black Sea Region with Details of Steamship Connections* (1844).

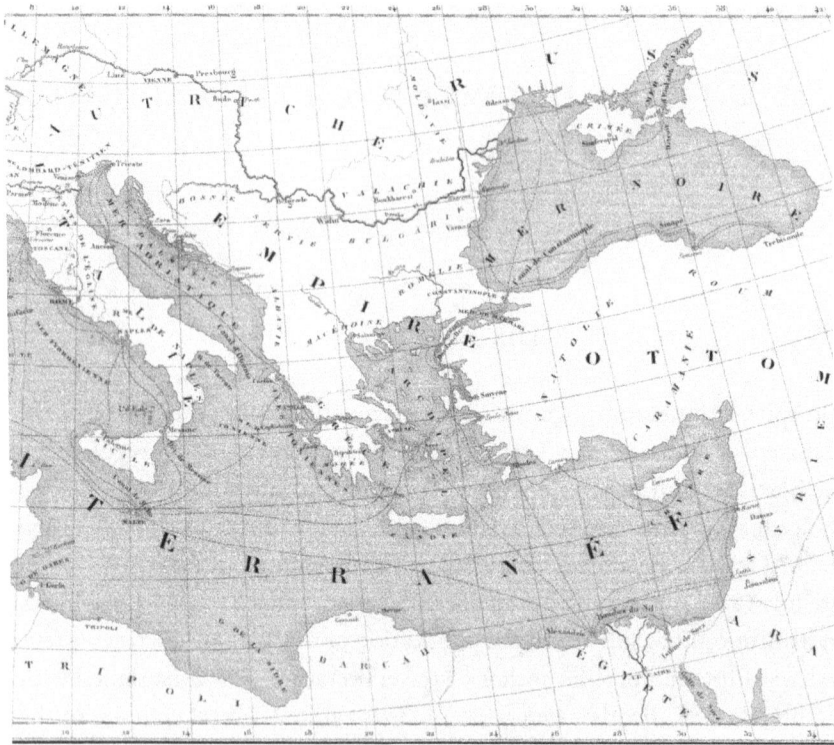

coaches per week, and the distance of around 210 km was covered in twenty-seven hours.[2] A journey by the direct overland route was equally demanding: the almost 1,000 km of road between Istanbul and Belgrade through the European part of the Ottoman Empire raised serious concerns, not just about comfort but also about safety. It is hardly surprising that the eastern route, by water, on board the vessels of the DDSG should seem much more attractive, simpler, and even more comfortable. A decade after their launch on the Habsburg Danube in 1830 and four years after they first managed to cover the entire route between Vienna and Istanbul, the Austrian steamers were carrying thousands of travelers annually between the two metropolises.

---

2   *A Handbook for Travellers in Southern Germany*, 2nd ed. (London, 1840), 334, 338, 350.

A number of English tourists and three French, a Belgian official, and a Polish adventurer numbered among Blewitt's traveling companions when he embarked on board the sea-going steamer *Ferdinand I* or *Ferdinando Primo* in Istanbul on July 30, 1840. After a stop at Varna, the vessel entered the Danube, stopping at Tulcea, Galați, and Brăila. Seen from its deck, the Danube ports had little to offer the passengers in the way of attractive land-scapes; they proved more interested in the beauty of some local women who came to look at the ship and at the foreigners on board. While they waited for the river steamer at Brăila, a dance, organized spontaneously, pro-vided a new occasion for mutual analysis and admiration between passen-gers and local women. The travelers then continued upstream on the river steamboat *Pannonia*, which stopped in the ports on the right ("Dobrogean," "Bulgarian," or "Ottoman") bank of the Danube. Later, they spent ten days of isolation in the quarantine station of Jupalnic (Orșova), at the entrance to Austrian territory, before continuing their journey toward Budapest and Vienna.[3]

From the mid-1830s, the Danube had become the central axis of an increasingly complex network of international transport that ensured the connection not only between two capitals but also between West and East. Through its function as both frontier and connector between empires, the Danube route was a special space in itself, neutral also from the point of view of international law, and one that promised travelers a fascinating voy-age through and between civilizations. The DDSG steamboats had man-aged to open to circulation an artery navigable for more than 2,000 km, a natural highway that had hitherto been fragmented and little used for the international transport of people and goods. With support from the govern-ment in Vienna, the DDSG extended its operations beyond the borders of the Habsburg monarchy, not only toward the Lower Danube and the Black Sea but also upstream, toward the German states. And, there was no short-age of customers, like Blewitt and his traveling companions, convinced that the Danube route was, at least in certain periods of the year, the best way of making the journey from Istanbul to Vienna.[4]

---

3   Brother Peregrine [Octavian Blewitt], "The Danube," part 1, *Fraser's Magazine for Town and Country* 22, no. 131 (1840), 560–572.
4   Details on the Danube route in Ardeleanu, "From Vienna to Constantinople on Board the Vessels of the Austrian Danube Steam-Navigation Company (1834–1842)," *Historical Yearbook* 6 (2009), 187–202.

From "a Journey of Toil and Danger, into One of Unmixed Pleasure and Enjoyment"

Richard T. Claridge (1797/9–1857) was one of the first British travelers who, at the end of a steamboat journey in 1836, compiled useful information into a travel guidebook that popularized the advantages of the new route for the British public. According to Claridge, the Austrian steamboats had "converted what was hitherto a journey of toil and danger, into one of unmixed pleasure and enjoyment"; in correlation with the other steamboat routes in the Black Sea, the Eastern Mediterranean, and the Adriatic, the Danube route offered the possibility of visiting some of the most fascinating destinations in Asia Minor, Greece, and Italy in return for a reasonable investment in time and money. In the second edition of his book, Claridge further expanded his list of West–East connections, giving details of routes between Paris and the Mediterranean, Paris and Munich, Munich and Trieste, and from London to the Rhine and on to the Danube, to Alexandria and Cairo, and to India.[5] The year of Claridge's first edition, 1837, also saw the publication in London of the second guidebook in the Murray's series (dedicated to travel in what we now call "Central Europe"), in which the Danube voyage between Vienna and Istanbul enjoys a detailed presentation.[6] A few years later, a German guidebook was published in Bavaria, including a description of the Lower Danube region,[7] a geographical space integrated, at least along the river, in the geography of international transport.

The Murray's series appeared in the context of an increasing interest in travel on the continent in a period of exponential growth in transport infrastructure and international tourism.[8] The editor, John Murray III (1808–1892), was quick to understand the change in routes and travel routines and to see the opportunity to publish volumes that promised reliable information about itineraries, transport services, hotels, and sights to visit. The full subtitle of the volume dedicated to travel in the south German space, which included

---

5   R. T. Claridge, *A Guide along the Danube* (London, 1837) and *A Guide Down the Danube* (London, 1839).
6   *A Handbook for Travellers in Southern Germany* (London, 1837).
7   Adalbert Müller, *Die Donau vom Ursprunge bis zu den Mündungen*, vol. 2, *Die untere Donau* (Regensburg, 1841).
8   Scott Laderman, "Guidebooks," in Carl Thomas (ed.), *The Routledge Companion to Travel Writing* (London, 2015), 278–288.

also the Habsburg territories, is abundantly clear: "Including descriptions of the most frequented baths and watering-places; the principal cities, their museums, picture galleries, etc.; the great high roads, and the most interesting and picturesque districts. Also, directions for travellers and hints for tours." The guidebooks were not to be compiled out of details gleaned from various sources, but had as their basis the direct experience of journeys made by the editors or their acquaintances. Readers were invited to send comments and "notices of any errors which they may detect," thus initiating the practice of updating such a work by collective effort.[9] The idea was quickly adapted for a German-speaking clientele by Karl Baedeker (1801–1859), a Koblenz publisher who had already produced travel guides for the Rhineland. Baedeker went further in standardizing the information, aiming to relieve the traveler of all concerns on the way. Guidebooks became indispensable articles in the luggage of any traveler. The French edition of one such volume, "indispensable for artists, merchants, and travelers," was also adapted in both its structure and its content to the new realities of the transport market, introducing information about the new routes through southeastern Europe. It thus included a table of post stations, railways, and steamboat routes, information about the means of travel specific to the various countries and about currencies and exchange rates, descriptions of towns and villages and of natural and artistic curiosities, and details about spas, trade, population, hotels, and so on.[10]

Seeking to come in support of an increasingly hurried clientele, these books not only provided clear information about the various travel options but also took on the role of summarizing what was important and worth seeing. Standardized details and short relevant anecdotes created an accessible narrative, often including generalizations loaded with the ideological and aesthetic convictions of the authors and of their time. In the German and Austrian space, such texts contributed to the promotion of cultural stereotypes that were influential in the construction of national and

---

9   Esther Allen, "'Money and Little Red Books': Romanticism, Tourism, and the Rise of the Guidebook," *Lit: Literature Interpretation Theory* 7, nos. 2–3 (1996), 213–226; Gráinne Goodwin and Gordon Johnston, "Guidebook Publishing in the Nineteenth Century: John Murray's Handbooks for Travellers," *Studies in Travel Writing* 17, no. 1 (2013), 43–61; Rebecca Butler, "'Can Any One Fancy Travellers without Murray's Universal Red Books'? Mariana Starke, John Murray and 1830s' Guidebook Culture," *Yearbook of English Studies* 48 (2018), 148–170.

10   J.-M.-V. Audin, *Guide classique du voyageur en Europe*, 2nd ed. (Paris, 1852).

supranational identities in the region.[11] The guidebooks to India analyzed by Kathleen Epelde, for example, include all the elements of European cultural hegemony specific to the Orientalizing discourse.[12] Ideas about identity and alterity were thus manifestly evident in these volumes, which were written from the perspective of Western authors convinced of the differences in civilization between Europe and the rest of the world.

All these antecedents of today's *Lonely Planet* or *Trip Advisor* also began to include in their guidebooks information about the exotic territories on the peripheries of the Ottoman Empire, including the autonomous states of Serbia, Wallachia, and Moldavia. In fact, these lay precisely along the Danube route, and some of the ports of the Principalities were important nodes in the functioning of the connection between the two imperial capitals, Vienna and Istanbul. The Danube route of the DDSG refreshed the pan-European transport corridors and accelerated the integration of the Principalities not only on the map of global mobility but also in the pages of influential travel guides.

ROADS, EMPIRES, CIVILIZATIONS

In many respects, Ami Boué (1794–1881) is a representative figure for the spirit of the nineteenth century, a century defined by nationalism and industrialization, but equally by internationalism and the sacralization of tradition. Born in Hamburg to French parents, Boué studied in Switzerland, France, and Scotland, where he received his training as a geologist at the University of Edinburgh. Boué traveled frequently throughout Europe. Settled in Vienna and naturalized as a citizen, he became a well-informed observer of the Balkan world, about which he wrote numerous works, including a monumental study of over 2,000 pages in which he presented "the geology, natural history, statistics, manners, costumes, archaeology, agriculture, industry, commerce, governance, clergy, history, and political situation" of the peoples of "European Turkey."[13]

---

11 Rudy Koshar, "'What Ought to Be Seen': Tourists' Guidebooks and National Identities in Modern Germany and Europe," *Journal of Contemporary History* 33, no. 3 (1998), 323–340; Josef Ploner, "Tourist Literature and the Ideological Grammar of Landscape in the Austrian Danube Valley, ca. 1870–1945," *Journal of Tourism History* 4, no. 3 (2012), 237–257.

12 Kathleen Epelde, *Travel Guidebooks to India: A Century and a Half of Orientalism*, PhD thesis, English Studies Program (University Wollongong, 2004), http://ro.uow.edu.au/theses/195 (accessed August 18, 2023).

13 Ami Boué, *La Turquie d'Europe*, 4 vols. (Paris, 1840).

In Boué's view, the "foreigner" who wanted to know the "Orient" had to travel by land.[14] On the maritime routes that converged in "Constantinople," the zero point of Ottoman civilization and the gate of the Orient, the traveler would encounter a number of coastal regions and cities before immersing himself completely in the "Asiatic world." On the land routes of the Balkans, the passage between the two cultural spaces of "Europe" and the "Orient" was much more gradual and subtler, and the experience of travel more profitable. The Habsburg Empire—or rather *Kaiserthum Oesterreich*—served as a necessary contact zone between "West" and "East," just as "European Turkey" conveyed by its very name the idea of an overlapping—in varying proportions—of the two "worlds." Vienna and Istanbul were the centers that marked one's complete "immersion" in the new "civilization." Between them stretched a space of contacts, influences, osmoses, a hybrid territory in which "the two worlds" interpenetrated in varying proportions and unique mixtures. The traveler who took the land routes between Vienna and Istanbul would thus pass, in the course of a journey of some 1,500 km, through geographically and culturally varied regions in which—at the level of material and spiritual life, in physiognomy and dress, in customs, manners, and the appearance of homes—the "Orient" gradually insinuated itself into the "Occident." A journey in the opposite direction was a veritable *anabasis*, a road toward "home," toward safety, toward civilization. In his fascinating study of the "inventing" of Eastern Europe, the American historian Larry Wolff emphasizes the special status of these transitional territories, "imagined" in the period of the Enlightenment as lying "between Europe and Asia." The information in the French *Encyclopédie* about Hungary and Bulgaria confirms this fluid geographic status,[15] in which "Western civilization" gradually gives way to a semi-Oriental barbarism.

Roads were the first element that indicated the differences between the two empires. The state of the public highways in Habsburg and Ottoman territories seemed to be governed by two opposing visions. In the former, transport infrastructure was of increasing economic and administrative relevance, connecting the various provinces of the monarchy and contributing to the well-being of its citizens. Enlightenment economic rationality made roads a vital component for the functioning of the imperial organism; internal regulations

---

14 Boué, *Recueil d'itinéraires dans la Turquie d'Europe*, vol. 1 (Vienne, 1854), 1–4.

15 Larry Wolff, *Inventing Eastern Europe: The Map of Civilization on the Mind of the Enlightenment* (Stanford, 2004), 183–186.

ensured the proper maintenance of the public road network and the functioning of a relatively efficient postal and diligence system, which facilitated the
mobility of travelers and goods the length and breadth of the empire. In the
Ottoman territories, especially in border provinces, the transport infrastructure was seen primarily in terms of its military and defensive function. Road-
building or maintenance works in territories on the other side of the border
were considered to be offensive actions. In 1816, for example, the news in
Banat that the roads were being repaired was considered the prelude to a fresh
conflict. In answer to the questions raised by the Wallachian authorities, the
Austrian consul in Bucharest, Fleischhackl von Hackenau, replied: "Austria
repairs old roads and builds new ones for the promotion of commerce and
internal circulation, which any state is entitled to do without being obliged to
give account to its neighbors."[16] Thinking in defensive terms, the Ottomans
seemed to privilege military roads and took special care of infrastructure that
had a strategic role in the defense of the empire. It should be added that we
may detect a change in attitude to public roads in the Principalities during the
Phanariot period, reflected in a gradual increase in the number of official documents referring to their repair.[17]

In one of his volumes, Boué presents seventeen overland routes along
which travelers circulated through "European Turkey," the antechamber of
the Orient. The Balkan Peninsula was crisscrossed by a complex network of
roads from Tulcea, Varna, and Burgas on its eastern edge to Shkodër and
Thessaloniki in the west and south.[18] The central axis of the Balkans followed the route between Belgrade and Istanbul and provided the most direct
overland connection between "Europe" and "Asia," along the line of the Via
Diagonalis or Via Militaris, the old Roman road through Belgrade, Niš,
Sofia, Plovdiv, Edirne, and Istanbul, with branches to the main military and
economic centers of European Turkey.[19] This was the route that travel guides

16 Andrei Oțetea, *Pătrunderea comerțului românesc în circuitul internațional (în perioada de trecere de la feudalism la capitalism)* (București, 1977), 141.
17 Laurențiu Rădvan (ed.), "Drumuri de țară și drumuri de oraș în Țara Românească în secolele XVII–XVIII," in *Orașul din spațiul românesc între Orient și* Occident (Iași, 2007), 68–69.
18 Boué, *Recueil d'itinéraires.*
19 Alexander Vezenkov, "Entangled Geographies of the Balkans: The Boundaries of the Region and the Limits of the Discipline," in Roumen Daskalov, Diana Mishkova, Tchavdar Marinov, and Alexander Vezenkov (eds.), *Entangled Histories of the Balkans*, vol. 4, *Concepts, Approaches, and (Self-)Representations* (Leiden, 2017), 215–225. See also the recent volume by Florian Riedler and Nenad Stefanov (eds.), *The Balkan Route: Historical Transformations from Via Militaris to Autoput* (Berlin, 2021).

published in the early nineteenth century recommended to travelers heading from Vienna toward Istanbul.[20]

According to the 1840 edition of Murray's guidebook, the distance between Constantinople and Belgrade could be covered by couriers in around a week, but for travelers, a journey of twelve days was recommended, including a day at Adrianople (Edirne) and another at Niš. In Ottoman territory, travel was by the postal service. One had to employ a Tatar courier, carry sufficient provisions, and accept primitive accommodation conditions.[21] One could tell when one had entered Habsburg territory not only from the better condition of the infrastructure but also from the superior organization of transport services. The postal service, which was under the imperial government, was considered to be efficient. There were numerous inns along the main highways, although those in Transylvania and Hungary were considered poorer and dirtier than in the western half of the monarchy. From Budapest, a diligence service offered a connection to Vienna along better roads and at a greater speed than in Bavaria. Customers were permitted to carry with them a piece of luggage of up to 10 kg, with the possibility of an additional chest of maximum 15 kg being carried in a separate luggage cart at no extra cost. Travelers could also make arrangements to travel by special diligences.[22]

To return to Boué, he insisted too little in his texts on the practical issues that a traveler would come up against along the roads of the Balkans. The idea of an overland pilgrimage through the "real" Oriental world was attractive for enthusiastic wanderers or adventurers eager to encounter unknown places and to enjoy unforgettable experiences. For the "ordinary" traveler, however, safety was the crucial aspect, and Ottoman roads were not considered to offer sufficient guarantees that travel would be safe, predictable, and comfortable.

THE WALLACHIAN CONNECTION

There were several possible variants on the direct route between Istanbul and Vienna—that is, by the Via Diagonalis and then through the Habsburg Empire. One of these went by the Shipka Pass, Vidin, and Požarevac before

---

20 Heinrich August Ottokar Reichard, *An Itinerary of Germany* (Paris, 1826), 502–504.
21 *A Handbook for Travellers in the Ionian Islands, Greece, Turkey, Asia Minor, and Constantinople* (London, 1840), 214–215.
22 Ibid., 126–127, 137–138.

arriving in Belgrade, which, from a geographical, political, and cultural point of view, was one of the symbolic points of meeting between empires. Another went across Wallachia, entering Transylvania by the valley of the River Olt. The principal nodes along this route were Burgas, Shumen, Ruse, Giurgiu, Bucharest, Pitești, Turnu Roșu, and Sibiu. The accounts of two British officers who chose this itinerary at the beginning of the second quarter of the nineteenth century offer an insight into the difficulty and insecurity of travel through the Ottoman territories and the improvement of travel conditions in the Habsburg provinces.

James Edward Alexander (1803–1885) crossed the Balkans in the autumn of 1826 at the end of a journey from India. He and his fellow travelers left Istanbul in the company of a Tatar courier, who was paid the substantial sum of 1,000 piastres (£20 pounds sterling). The group traveled continuously through a wild but picturesque region. Alexander mentions the number of boars, bears, wolves, and deer they encountered on their way and the fact that he "sabred a black snake, six feet in length, which possessed the poisonous fangs." One of the servants engaged in Istanbul, unused to the demanding rhythm of the journey, suffered a panic attack and accused the Tatar courier of wanting to kill him. In five days, the travelers reached Ruse, and as there was an outbreak of plague there, they hastened to cross the Danube. They traveled from Giurgiu to Bucharest in extremely uncomfortable conditions, in "low waggons, each containing a single person" drawn by four horses that raced across the plain. After spending a few days in the Wallachian capital, the Britons set out for Sibiu in the same "accursed vehicles" or, to ease their shaken bones, on horseback. After entering the Austrian monarchy via the quarantine station at Turnu Roșu and spending a few days in Sibiu, they continued their journey, "travelling in an excellent carriage and four through Transylvania, Hungary, and Austria." After another five days on the road, they arrived in Vienna.[23]

Another British officer, Charles Colville Frankland (1792–1876), traveled in the opposite direction, passing through Timișoara, Lugoj, Deva, and Sibiu in late March of 1827. He describes in rich detail his journey along the better-quality roads of Hungary and the Banat. Difficulties arose when he encountered a broken bridge near Lugoj and his carriage had to be taken across "a deep brook" (the Timiș), and when horses had to be changed at post stations. Already in

---

23   James Edward Alexander, *Travels from India to England ... etc. in the Years 1825–26* (London, 1827), 242–257.

Transylvania the roads were getting worse and worse, and in Wallachia "they became dreadful, full of immense stones and deep ruts." The passage through the gorge of the Olt was a torment. On March 31, a day of "threatening and gloomy" weather, the travelers crossed the Olt in a "flat boat" and "disembarked on a rocky and uneven shore, full of streams and ravines." They frequently had to deviate from the road, and in places got out of their carriage and walked:

> The roads or tracts became more and more difficult and steep as we advanced; often leading over ravines which are traversed by means of rotten and trembling platforms of trunks of trees, so ill put together that frequently the horses ran the greatest risk of breaking their legs, by falling into the spaces between tree and tree. Indeed it is not easy to imagine any thing more arduous than these ascents, or more determined and persevering than the Wallachian postilions and guides; for I am convinced that no civilized man would entertain an idea of the practicability of these passes.

Fresh accidents punctuated the travelers' journey toward Bucharest. At one point, they had to supplement the eight horses of their carriage with an additional four from their baggage wagon in order to get over a particularly steep and precipitous stretch of road. The crossing of the swollen Argeş River proved to be a fresh challenge, as did the journey to the Wallachian capital, along roads "very much cut up by the rains, and full of deep ruts and bogs." From Bucharest, the Britons set off for the Danube in low carts:

> These little waggons are about a foot and a half from the ground, and are the rudest and most extraordinary vehicles I have ever yet seen or heard of. You are dragged along, with immense rapidity, through bogs and ruts, over brushwood, and through ravines and streams, seated upon a truss of hay, and nearly shaken to death by the violence and rapidity of the motion.

From Ruse, they continued on horseback toward Istanbul, where they arrived, via Shumen and Burgas, almost a week later. The journey was not without adventures, from the crossing of swollen rivers to disputes over accommodation in inns and the procurement of horses in post stations.[24]

––––––––––

24 Charles Colville Frankland, *Travels to and from Constantinople, in the Years 1827 and* 1828 (London, 1829), 1–92.

Both descriptions are illustrative of travel conditions in central and southeastern Europe, at least from the perspective of Western writers. If in the Habsburg Empire the state of the roads was relatively good, and the organization of postal services allowed one to travel in decent and predictable conditions, in the Ottoman territories (including the Danubian provinces subordinated to the Porte) the roads were almost impassable for much of the year. The discomfort created by the poor quality of the road infrastructure was increased by the lack of bridges, which were particularly necessary in the spring, the period of high water, and the winter, when rivers and streams had to be forded. The postal service was also deficient, which made travel unpredictable and risky. A journey between Vienna and Istanbul across the Balkans lasted two to three weeks and, in addition to readiness to undertake a prolonged effort, required considerable expense.

Despite these problems, merchants, administrators, and diplomats circulated frequently along these routes.[25] However, any journey necessitated thorough preparation. In other words, despite the encouragement of writers such as Boué, relatively few travelers chose to circulate on the roads of the Balkans, and then only when they had no better option. The introduction of steam navigation on the Danube created just such an option.

GONE WITH THE WIND

Before turning to the context in which the Austrian steamers appeared, something should be said about the sailing vessels circulating between Istanbul and the ports of the Lower Danube. For the period before 1830, statistical information is relatively scanty, but it is clear that hundreds of small vessels were sailing along this route every year. For reasons of safety and comfort, plenty of travelers considered them the best solution for a journey between Istanbul and the Danube ports, or at least for part of the route. Around 1815, the young Dimitrie Foti Merişescu traveled to Istanbul in the suite of the Wallachian *capuchehaia* (diplomatic representative) Dimitrie Caragea, via Brăila, Măcin, Hârşova, Varna, and Nesebar, where "we

---

25 Anon., "The Different Roads from the Lower Danube to Constantinople," *Morning Chronicle*, London, July 30, 1828.

boarded a ship and went to Tsarigrad [Istanbul]."[26] On August 12, 1822, Ioniţă Sandu Sturdza, the newly appointed ruler of Moldavia, set out on the Black Sea on his return from Istanbul, "together with his suite, not to mention two Moldavian boyars." These latter, "being afraid to go by ship on the sea," went overland in the suite of the Wallachian ruler Grigore Dimitrie Ghica. Sturdza had chosen the sea route, not only because it was cheaper but also because of his precarious state of health, which did not allow him to make the tiring journey over the Balkan Mountains. On board the vessel, the travelers relieved their boredom by playing stoş,[27] and to maintain their rank, they gave charity to "poor Turkish women." In a few days, they reached Varna, from where they went overland to Silistra and on to Iaşi.[28] The German doctor F. S. Chrismar traveled from Munich to Istanbul in 1833, passing through Austria, Hungary, Transylvania, and Wallachia. He embarked at Galaţi on board the sailing vessel *Virgin of Hydra*, hoping that in the company of its experienced Greek crew he would be better placed to face the voyage on the "perfidious" Black Sea. Chrismar tells how the proud European river gradually broke up into the Russian-controlled marshes of the Delta, making the journey all the more difficult: "We had to spend two days between these measureless marshes, as we could only go short distances, due to the slow flow of the Danube and the total lack of a favorable wind." The ship reached Sulina on the sixth day since its departure from Galaţi, and, after a further five days floating on the Black Sea, the passengers at last caught sight of the light at the mouth of the Bosphorus, the border between Europe and Asia.[29]

The great problem of sea voyages was their unpredictability. According to the calculations of the Greek historian Apostolos Delis, a sailing ship took on average nineteen days to get from Istanbul to Galaţi, compared with just six days to get to Odesa; the difference underlines the complications of navigation in the region of the Danube Delta, where the course of the river was sinuous and often blocked by sandbanks. In the cold season, when the river

---

26 Constanţa Vintilă-Ghiţulescu, *Tinereţile unui ciocoiaş. Viaţa lui Dimitrie Foti Merişescu de la Colentina scrisă de el însuşi la 1817* (Bucureşti, 2019), 111–112.

27 A card game similar to Faro.

28 Petronel Zahariuc, "Bacşişuri, mătăsuri şi argintării. Călătoria boierilor moldoveni la Constantinopol în 1822," in Dan Dumitru Iacob (ed.), *Avere, prestigiu şi cultură materială în surse patrimoniale. Inventare de averi din secolele XVI–XIX* (Iaşi, 2015), 323–324.

29 F. S. Chrismar, *Skizzen einer Reise durch Ungarn in die Turkei* (Pest, 1834), 125–137 (quote at 136).

was not frozen, the journey took longer and was more dangerous.[30] Steam-powered vessels promised to solve at least some of these problems and to bring navigation into a new age of predictability, safety, and comfort.

STEAMBOATS ON THE DANUBE: CAPITALISM, ENTREPRENEURSHIP, AND IMPERIAL HYDROPOLITICS

The introduction of steamboats on the Habsburg Danube and the extension of the service to the Lower Danube and the Black Sea resulted from the convergence of three main factors, which I propose to examine in the following pages: the desire for profit on the part of private entrepreneurs interested in making money from a technological invention; the involvement of enlightened administrators seeking to instrumentalize the transport revolution for the material progress of peripheral areas of the Austrian monarchy; and the importance of the navigation service for the strategic interests of the government in Vienna against the background of the deepening crisis of the Ottoman Empire (the so-called Eastern Question).

In January 1829, British investors Joseph Prichard and John Andrews launched in Vienna their project of bringing steam vessels to the Danube. The initiative aroused the interest of a number of important bankers and leading politicians, and the two hundred shares of 500 florins each were rapidly subscribed. Armed with an authorization obtained from the imperial authorities, the entrepreneurs proceeded to build their first steamboat in the Erdberg shipyard, near Vienna. It was equipped with a 60 hp steam engine supplied by Boulton & Watt, one of the most renowned British manufacturers in the field. The vessel, named *Franz I* in honor of the emperor, and the company, the DDSG, would make history.[31]

For more than two decades, steamboats had been revolutionizing the transport of passengers and goods in North America and Western Europe. On the Hudson, Mississippi, and Ohio, on the Clyde, Thames, and Rhine, on the seas and oceans of the world, steam-powered vessels proclaimed the

---

30 Apostolos Delis, "Navigating Perilous Waters: Routes and Hazards of the Voyages to Black Sea in the Nineteenth Century," in Maria Christina Chatziioannou and Delis (eds.), *Linkages of the Black Sea with the West. Navigation, Trade and Immigration* (Rethymnon, 2020), 20–21.
31 A history of the company in Erste Donau-Dampfschiffahrts-Gesellschaft, *125 Jahre Erste Donaudampf-schiffahrtsgesellschaft* (Wien, 1954).

independence of mobility from the rhythms and whims of nature. Grandiose projects were underway everywhere, and Vienna was no exception. Here, as early as 1817, a number of investors had toyed with the idea of introducing steamboats on the Danube, but, for financial or technical reasons, their plans never materialized. Prichard, who had experience in the great Woolwich shipyard (on the Thames, close to London), and Andrews, who knew German[32] and was probably well acquainted with Austrian business circles, managed to mobilize the necessary financial and human resources for the construction and equipping of the steamboat *Franz I*. Its maiden voyage in September 1830 was a resounding success: the distance from Vienna to Budapest was covered in fourteen and a half hours downstream and forty-eight and a half hours upstream.[33]

The transport revolution had arrived in the heart of the Habsburg Empire, making possible its economic development along its spinal column, the Danube. From February 1831, a regular service was introduced on the Vienna–Budapest route, with a seasonal continuation as far as the entry to the Iron Gates gorge. However, the company had plans to invest in the construction and equipping of new steamboats with which to pass beyond the gorge and establish a continuous connection between Vienna and Istanbul via the Black Sea.[34]

It took more than the brute force of steam to conquer the great river: human energy was needed too. From this point of view, the support of a Hungarian nobleman, Count Széchenyi István (1792–1860), proved decisive for the success of the Viennese steamboat company.[35] Appointed royal commissioner for the development of Danube navigation, Széchenyi, himself a

---

32 Anon., "Introduction of Steam Navigation into Austria," *Mechanics Magazine, Museum, Register, Journal and Gazette* 48 (1848), 103.

33 Henry Hajnal, *The Danube: Its Historical, Political and Economic Importance* (The Hague, 1920), 124–125.

34 Anon., "Die erste k. k. pr. Donau-Dampfschiffahrt-Gesellschaft. Erste Periode. Gründung und Betriebs-Verhältnisse bis zum Jahre 1841," *Centralblatt für Eisenbahn und Dampfschifffahrt in Österreich* 35 (August 30, 1862), 337–340; "Zweite Periode. Vom Jahre 1842–1851," *Centralblatt für Eisenbahn und Dampfschifffahrt in Österreich* 36 (September 6, 1862), 345–348; 37 (September 13, 1862), 353–355; and 40 (October 4, 1862), 377–379. Other general presentations: *Denkschrift der ersten k.k. privilegirten Donau-Dampfschiffahrts-Gesellschaft zur Erinnerung ihres fünfzigjährigen Bestandes* (Wien, 1881); Helmuth Grössing et al., *Rot-Weiss-Rot auf Blauen Wellen* (Wien, 1979); and Franz Dosch, *180 Jahre Donau-Dampfschiffahrts-Gesellschaft* (Erfurt, 2009).

35 Miklos Szucs Nicolson, "Count Istvan Széchenyi (1792–1860): His Role in the Economic Development of the Danube Basin," *Explorations in Economic History* 6, no. 3 (1954), 163–180.

DDSG shareholder and an enthusiast for modern entrepreneurial ideas, collaborated closely with the representatives of the company in the coordination of the engineering works at the Iron Gates (to be discussed at greater length in Chapter 6), but especially in setting up the institutional framework that permitted navigation beyond the borders of the empire. Travel agencies had to be set up, piers built, and stores to supply coal established.[36] Money was invested in mapping the river, in recruiting and training sailors, in marketing, and so on. New steamboats were built and launched: *Argo* (1833), *Pannonia* (1834), *Maria-Dorothea* (1834), *Zrinyi* (1835), *Ferdinand I* (1836), and so on. In April 1834, the steamboat *Argo* passed through the Iron Gates, symbolically uniting the Middle and Lower Danube. It thus became possible to extend the DDSG service to the great ports of the Maritime Danube—Brăila and Galați—and to connect Vienna, via the Black Sea, to Istanbul. Sea-going steamers were built, and the service became fully functional in 1836, when the *Ferdinand I* began to operate on the Galați–Istanbul segment.[37]

The international political context favored such initiatives, as interest in the Eastern Question had grown exponentially. In 1833, Russia obtained important advantages by "saving" Sultan Mahmud II (1785–1839, reigned 1808–1839), who was threatened by the ambitions of his own vassal Mehmet Ali, the pasha of Egypt (1769–1849); and the great powers, led by Great Britain and Austria, took diplomatic action to return to the period of strategic equilibrium in the region of the Bosphorus and the Dardanelles. With the eyes of European diplomacy focused on the Sublime Porte, the project of linking Vienna to Istanbul by way of the Danube and the Black Sea acquired a particular geopolitical relevance and was more and more encouraged by the political leaders in Vienna. As early as the DDSG shareholders' meeting of January 1834, Baron Ottenfels, the imperial internuncio (i.e., the Austrian ambassador to Istanbul), made known the government's interest in the company and in navigation on the Danube, the Black Sea, and the

---

36 Virginia Paskaleva, "Le rôle de la navigation à vapeur sur le bas Danube dans rétablissement de liens entre l'Europe Centrale et Constantinople jusqu'à la guerre de Crimée," *Bulgarian Historical Review* 4, no, 1 (1976), 64–76; Paskaleva, "Shipping and Trade on the Lower Danube in the Eighteenth and Nineteenth Centuries," in Apostolos E. Vacalopoulos, Constantinos D. Svolopoulos, and Béla K. Király (eds.), *Southeast European Maritime Commerce and Naval Policies from the Mid-Eighteenth Century to 1914* (Boulder, 1988), 131–151.

37 See the larger context in Charles King, *The Black Sea: A History* (Oxford, 2004).

Eastern Mediterranean. Chancellor Metternich was a supporter of hydropolitical projects, and the DDSG's privilege was repeatedly prolonged.[38]

From 1845, the route between Galați and Istanbul was taken over by the Austrian Lloyd shipping company, founded at Trieste in 1833. The DDSG remained a river navigation company, while Austrian Lloyd operated maritime steamer routes in the Black Sea and the Eastern Mediterranean. The agreement between the two private companies was reached with the involvement of the government in Vienna, which insisted on the political and economic importance of maintaining Austrian control of these interimperial routes. At the same time the government took an increasingly close interest in the DDSG. From 1843, it was placed under the control of the Habsburg chancery, and in 1846, it was granted the exclusive right to sail on the Danube, in exchange for carrying postal correspondence on its steamboats.[39] Up until the First World War, the DDSG enjoyed significant fiscal privileges and represented not only the financial interests of its shareholders but also the political and economic interests of the government. These observations regarding the government support given to the DDSG apply also to the Austrian Lloyd company, whose steamers connected Trieste to the great ports of the Eastern Mediterranean and the Black Sea.

Organization, Travel Times, and Costs

The achievement of a river and sea connection between Vienna and Istanbul was a long process. For it to be completely operational required not only the equipping of the DDSG with steamboats for the various segments of the route but also the administrative and logistic organization mentioned above.

The journey between Vienna and Istanbul consisted of four distinct segments, two on either side of the Iron Gates. Passage through the gorge required special arrangements, which will be discussed in Chapter 6. Table 1 presents passenger figures for each of the four segments in 1843 and 1844. The busiest and most profitable was that between the great urban centers of the Habsburg monarchy—Vienna and Budapest—with stops at other important towns such

---

38 Hajnal, *The Danube*, 126–136.

39 Ronald E. Coons, *Steamships, Statesmen, and Bureaucrats. Austrian Policy towards the Steam Navigation Company of the Austrian Lloyd 1836–1848* (Wiesbaden, 1975), 108–114. Details about the early history of the company in *Die Dampfschiffahrt-Gesellschaft des Oesterreichisch-Ungarischen Lloyd von ihrem Entstehen bis auf unsere Tage 1836–1886* (Trieste, 1886), 4–43.

as Bratislava and Esztergom. The second internal segment, between Budapest and Drencova, was also relatively busy. Sailings were less frequent and passengers fewer on the following two segments, with more and more passengers traveling third class—in other words, on the deck. In the following pages, we shall look in more detail at the stages of the voyage, with an emphasis on the two segments covering the present-day Romanian space.

Table 1 Number of passengers on board the DDSG steamers

Year	1843				1844			
Class	1	2	3	Total	1	2	3	Total
Vienna–Budapest	54,775	77,200	10,345	142,320	59,146	83,371	17,514	160,031
Budapest–Drencova	11,626	20,150	2,262	34,038	10,843	18,725	4,882	34,450
Orșova–Galați	808	837	2,093	3,738	818	912	2,373	4,103
Galați–Istanbul	391	363	4,289	5,043	294	379	5,867	6,540

Source: Compiled from *Bilanz samt den dazu gehörigen Rechnungs-Ausweisen der österr. k. k. priv. ersten Donau-Dampfschiffahrts-Gesellschaft für das Jahr 1843* and *Bilanz ... für das Jahr 1844* (Wien, 1844 and 1845), unnumbered pages.

Theoretically, the schedule of sailings was established in such a way that passengers would be able to complete the whole journey as fast as possible. In practice, however, there were frequent delays. In the first years after the opening of the service, the journey took thirteen to seventeen days downstream and seventeen to twenty days upstream, not counting the period of isolation in the lazaretto on entering Habsburg territory. In subsequent years, the DDSG invested heavily in improving traveling times. With the introduction of more powerful steamboats and simplification of the administrative arrangements, the duration of the voyage was considerably reduced. For a few years, in addition to the usual route by river, the DDSG also made use of an alternative route across Dobrogea, with passengers disembarking at Cernavodă and being transported overland to Constanța.[40] The transfer

---

40 The National Archives of the United Kingdom, Foreign Office (hereafter TNA, FO) 78/363, fol. 28–30 (R. G. Colquhoun to Palmerston, Bucharest, April 12, 1839).

reduced the average journey time by two days. In the 1850s, there was clear progress in this respect, with express steamers completing the voyage in a week downstream and around ten days upstream.[41] By popularizing sailings that lasted around a hundred hours between Vienna and Galați and another fifty from Galați to Istanbul,[42] the DDSG tried to reposition itself on an increasingly competitive market, in which, at least within the Austrian territories, it had to hold its own against the faster and more comfortable railways. With a good connection between Galați and Odesa, the company's management banked on increasing demand in the Ukrainian provinces in the south of the Russian Empire, a region where the terrestrial transport infrastructure was still insufficiently developed.

Prices were fixed according to a table that took into account the distance traveled and the class of comfort (see Table 2). For some segments, there was also a small difference between the price of travel upstream and downstream. Children under the age of ten paid half-price, and each traveler was allowed around 30 kg luggage, with extra charges for additional weight.[43] Meals were not included in the price of the ticket, but restaurant services were available on board the steamers (see Chapter 2).

Table 2 Travel cost (upstream) on various segments (in florins and kreuzers)

Route/class	1	2	3
Istanbul–Galați	55.00	40.00	15.00
Galați–Schela Cladovei	35.00	24.30	14.00
Drencova–Budapest	17.00	11.20	–
Budapest–Vienna	9.00	6.00	–

Source: Compiled from [Jean-Baptiste] Marchebeus, *Voyage de Paris à Constantinople par bateau à vapeur* (Paris, 1839), 281–286.

In every respect, journeys on the DDSG steamers exemplify what historians have termed the commodification of travel, a process that took place in the Western world in the first half of the nineteenth century. Some authors

---

41 Details in Ardeleanu, *International Trade*, 25–26.
42 Constantin Bușe, *Comerțul exterior prin Galați sub regim de port-franc (1837–1883)* (București, 1976), 96.
43 Details on prices and conditions in [Jean-Baptiste] Marchebeus, *Voyage de Paris à Constantinople par bateau à vapeur* (Paris, 1839), 281–286.

have insisted on the revolution in transport, examining the technical and economic progress that made journeys faster, safer, and cheaper, while others have focused on the radical changes in the experience of travel as a journey became a "service" to be purchased over a counter. While, before the appearance of companies like the DDSG, travel was "produced" by the traveler, who used his knowledge and financial resources to obtain the necessary means of transport and provisions for his journey (as Alexander and Frankland had done by negotiation), later all this was reduced to the purchase of a ticket that offered the passenger a "finite product": simple, safe, fast, and comfortable carriage to the destination.[44] For ports such as Giurgiu, Brăila, or Galați, in a peripheral space at the beginnings of its capitalist development, the arrival of such a modern service was even more important than in Vienna or Budapest. It would be followed by the commodification of other public services in the Principalities, including the introduction of travel by diligence.

INTERNAL ROUTES ON THE HABSBURG DANUBE

The Irish journalist Michael J. Quin (1796–1843) was the author of the first "bestseller" about the experience of a voyage on board the DDSG steamboats. In the late summer of 1834, Quin was preparing to travel from Paris to Istanbul "by the ordinary and very fatiguing course overland through Vienna, Semlin, and Belgrade," when he learned that the steamboat service had been extended to the Lower Danube. It was a new itinerary, attractive and convenient, so he headed for the Habsburg monarchy and embarked on the Austrian steamboat at Budapest on the night of September 24, 1834. The vessel left at 7 a.m., three or four hours later than scheduled, and the voyage was to meet with all sorts of obstructions. The Danube had numerous meanders, and such obstacles as water mills, local people's boats, and a bridge of boats connecting the towns of Novi Sad and Petrovaradin[45] were sources of considerable inconvenience to navigation. In Quin's opinion, the steamboat's English captain, Cozier by name, was "little conversant with any branch of nautical science" and seemed unfamiliar with the hydrography of the Danube. On several occasions, the vessel touched the river bed, "very much to

44 Will Mackintosh, "'Ticketed Through': The Commodification of Travel in the Nineteenth Century," *Journal of the Early Republic* 32, no. 1 (2012), 61–89.
45 Today part of Novi Sad.

the captain's astonishment and perplexity." On one occasion, the goods carried on board had to be unloaded in order to release it from a sandbank. As the steamboat was anchored at night to avoid accidents, it was September 29, five days after leaving Budapest but before the passengers reached Moldova Nouă, at the entrance to the Iron Gates gorge.[46]

Quin's book, published in 1835 and immediately translated into French and German, is illustrative of travel conditions in the early period of steam river navigation. To make circulation on the Danube possible, the government in Vienna and the DDSG directed their attention to various aspects, from hydraulic works and cartography to the training of human resources in the nautical field. Gradually, the Danube was transformed into a veritable spinal column of the Habsburg monarchy, an element of infrastructure vital for the economic life of the empire and an important part of its identity.[47] Already in the 1830s, the DDSG steamboats ensured the consolidation of these functions.

As mentioned above, the stretch of the Danube on the way from Vienna to Istanbul consisted of two segments. The first connected the great urban centers of the monarchy—Vienna and Budapest—via Bratislava, while the second connected Budapest to the Iron Gates gorge, at the furthest reaches of the empire's military frontier. In the first years after the introduction of steam navigation, the journey between Vienna and Budapest was made in one or two stages, depending on the situation of the navigable channel on the sector upstream from Bratislava. When steamboats were unable to reach the Vienna Prater, a diligence circulated regularly between Vienna and Bratislava, covering the distance of some 80 km in around seven hours.[48] It was also possible to go by light carriage along the post roads or by boat on the Danube. The journey downstream between Vienna and Budapest took approximately twenty-four hours, and the return journey, upstream, around twice as long. The times varied according to the weather conditions and the season, which had a great influence on the power of the current. Long delays were caused relatively frequently by the navigation rules, which obliged captains to anchor their vessels at nightfall or in foggy conditions, but allowed

---

46 Michael J. Quin, *A Steam Voyage Down the Danube: With Sketches of Hungary, Wallachia, Servia, Turkey, etc.*, 2nd ed., vol. 1 (London, 1835), 1–84 (quotes at 1, 11, 20).
47 Király, *"Die Donau ist die Form,"* 18–21.
48 Claridge, *A Guide along the Danube*, 29.

them to continue their journey during long summer evenings or on nights with a full moon. In the summer of 1842, steamboats circulated daily between Vienna and Budapest, stopping at Bratislava. The journey took three hours to Bratislava and another ten to Pest. Twice a week, there was a direct service between the two cities. It took thirty-four hours upstream from Budapest to Bratislava and another thirteen from there to Vienna.[49]

In the first decade of the DDSG's activity, the steamboats *Franz I, Nador, Galatea, Árpád,* and *Maria Anna* were in circulation on the Vienna/ Bratislava–Budapest sector. Later, in the 1850s, the number of vessels increased, and there were daily sailings between the imperial capital and Budapest, covering the distance in twelve to thirteen hours downstream and twenty-five hours on the return journey against the current. After the completion of the railway connection between the two cities, guidebooks recommended that visitors take the Danube route downstream to enjoy the picturesque views it offered and return by train. The railway journey from Budapest lasted eleven hours at the most.[50]

The testimony of the Moldavian monk Chiriac, mentioned in the Introduction, is interesting with regard to traveling conditions in the region in the early 1850s. After traveling to Kyiv, Chiriac visited Voronezh, Moscow, St. Petersburg, and Warsaw before returning "through Russia, Lehia [Poland], Nemția [Germany], Hungary, and Transylvania to the monastery of Secu." From Vienna, he set out by train, happy that in the compartment he met "some merchants speaking Moldavian." For part of his journey, he took the Danube steamer. The scholarly monk explains the difficulties of the cruise:

> And, as there was terrible fog, it obeyed the necessity for us to further tarry, and after the fog began to rise, it set off. But going downstream about half an hour and seeing that in that direction the fog had not yet lifted, it returned again to its place. And further tarrying more than an hour, it set off, for now the fog had risen. And I wondered why they do not go when there is fog, and together the travelers told me it is because the Danube is shallow and spread out and has many sandbanks in its

---

49 *A Handbook for Travellers in Southern Germany,* 3rd ed. (London, 1844), 425.
50 *A Handbook for Travellers in Southern Germany,* 7th ed. (London, 1857), 498.

middle. For that they are afraid to go in the fog, not to get lost, and moreover, if it ends up in some sandy place, it will be in danger. And indeed this was the reason, for now it was bright and it was possible to see, but in two places it scraped its bottom on the sand, so much so that everyone was scared.[51]

The second internal segment of the route from Vienna to Istanbul was that between Budapest and the Iron Gates gorge, initially served by the steamboats *Zrinyi*, *Franz I*, and *Galatea*. The vessel raised anchor early in the morning and the journey downstream took two to three days, particularly because the route was much traveled and the steamboats stopped in intermediary ports to pick up or set down passengers and goods. In 1836, there were two sailings per month, covering the downstream journey to Drencova in around two and a half days in the summer and up to four days in the remainder of the year.[52] Later, in 1842, steamboats left Budapest for Drencova twice a week.[53] When the water level was low, captains were obliged to stop at Moldova Nouă, a village some 18 km upstream from Drencova, as sandbanks obstructed navigation in the area. Beyond Drencova, rocks rendered it completely impossible.[54]

In the 1850s, there were five sailings per week along the busy stretch from Budapest to Zemun/Semlin (now a district of Belgrade), covering the distance in around thirty-two hours downstream and thirty-nine upstream. Between Zemun and Orşova, there were two sailings per week, of fourteen hours downstream and twenty upstream. The Iron Gates gorge, long considered impassable, could now be sailed with ease and in safety. When the water level was high, the steamers from Budapest continued as far as Schela Cladovei (in the present-day city of Drobeta-Turnu Severin); when it was very low, they stopped at Drencova. At such times, passengers would be transferred at Drencova or Orşova to a shallow-draft vessel, so that there would be as little interruption as possible along the route from Vienna to Istanbul. I shall return to the adventures of travel through this picturesque region in Chapter 6.

---

51 Giuglea, "Călătoriile călugărului Chiriac," 706.
52 *A Handbook* (1837), 367.
53 *A Handbook* (1844), 439.
54 Adolphus Slade, *Travels in Germany and Russia, Including a Steam Voyage by the Danube and the Euxine from Vienna to Constantinople, in 1838–39* (London, 1840), 159.

Projects to extend steam navigation on the river continued upstream too. In 1838, the DDSG and a Bavarian and Württemberg company introduced regular sailings between Vienna and Regensburg. The German company also covered the route as far as Linz and later, depending on the depth of the river, on to Ulm. The steamboats were linked to the schedule of diligences and trains in the German space, ensuring that travelers could be transferred for destinations in Western Europe.

The DDSG fleet was continually growing throughout this period. In 1840, Ami Boué noted that the company had thirteen steamboats, including two built of iron (one of 60 hp and the other of 76 hp) and a tug downstream from Budapest.[55] By 1850, the DDSG fleet had expanded to twenty-five vessels: five tugs and twenty steamboats of between 36 hp and 110 hp for the transport of travelers and goods.[56] In 1857, the DDSG had 87 steamboats and tugs, 270 cargo vessels, and other boats for transport,[57] making it the largest river navigation company in the world.

Although it extended its activities beyond the Iron Gates, the bulk of the DDSG's business was concentrated in the Habsburg territories. To return to the metaphor of circulation, the river was the aorta connecting the heart of the monarchy with the rest of the imperial organism, which received the oxygen necessary for economic life especially by way of the Danube steamboats, the red blood cells of the empire.

ALLA TURCA, ALLA ROMAIKA

A significant challenge for travelers along the Lower Danube, once they had emerged from the Danube Gorge, was posed by the system of quarantine instituted under Russian supervision in Wallachia and Moldavia against infection from the southern, Bulgarian, or Ottoman bank of the river (see Chapter 5). The DDSG's solution was to use two steamboats on this segment, each serving the ports on one bank only.

The Swiss William Rey (1821–1888) was a passionate traveler throughout the world. A graduate of the Vienna Polytechnic, he taught physics and

---

55 Boué, *La Turquie d'Europe*, 153.
56 Adolphe Joanne, *Voyage en Orient*, vol. 1 (Bruxelles, 1850), 17.
57 *A Handbook* (1857), 496. Information on the fleet in Anon., "Steam Navigation of the Danube," *The Era*, London, January 23, 1853.

mathematics and later became a busy entrepreneur involved in the insurance business in Italy, France, and Switzerland.[58] In 1848, Rey made a journey down the Danube, from Vienna to Istanbul. He recalled how, on the way downstream from Schela Cladovei, the pilot of the steamer made use of geographical indications to maneuver the vessel: instead of directing the helmsman to steer *right* or *left*, his commands were "alla Turca" and "alla Romaïka," meaning toward the "Turkish" or the "Romanian" (Wallachian) side, respectively.[59]

Rey's observation aptly illustrates the operation of the Austrian steamboats, each of which served only one bank of the river at a time. The "Wallachia Line" started from Schela Cladovei and stopped, depending on the passengers and goods, at Calafat, Islaz, Giurgiu, Oltenița, Brăila, and Galați, where there was a transfer to the sea-going vessel that completed the journey to Istanbul. The "Turkey Line" started from the village of Kladovo in Serbia (across the river from Schela Cladovei) and stopped at Vidin, Lom Palanka, Nikopol, Svishtov, Ruse, Tutrakan, Silistra, and Hârșova. For a number of years passengers also had the option of disembarking at Cernavodă. They would then cross Dobrogea overland and reembark on the Black Sea steamer at Constanța, thus reducing the journey by two days.

The distance from Schela Cladovei to Galați was covered in three or four days, depending on the weather, and subject to the frequent grounding of the steamers on sandbanks. The weight of the overloaded vessels slowed down the journey, as did long stops at the ports along the way. In 1842, there were six departures per month from Schela Cladovei, three *alla Romaïka* and three *alla Turca*, with the possibility of the overland shortcut to Constanța for passengers heading to Istanbul. In the 1850s, there were weekly sailings between Orșova and Galați, the journey taking four and a half days for a "normal" steamer.[60] The vessels that operated in this segment included the *Argo*, the *Pannonia*, the *Zrinyi*, the *Árpád*, the *Boreas*, the *Pest*, the *Széchenyi*, and the *Prince Metternich*. There was also, as mentioned above, an express steamer service from Vienna to Galați.

In the context of the integration of the trade in Danubian cereals in global markets starting from the 1830s, the Danube towns flourished as never before, with the river serving as the commercial axis of the two Principalities. Recently, the architect Ilinca Păun-Constantinescu has studied the decline and contraction of some of the Danube towns on the Romanian side,[61] a peripheralization phenomenon connected not only to the economic transformations of the post-communist period but also to the gradual loss of the Danube's function as a commercial highway for the Romanian space. The Austrian steamboats that circulated on the Lower Danube contributed directly to urban development in the Principalities (and in Bulgaria), with the Romanian river ports of Turnu Severin, Giurgiu, Brăila, and Galați turning into nodes of international and regional transport corridors precisely in the period that interests us in this book.

The genesis of the modern city of Turnu Severin can be traced in the period after the Treaty of Adrianople, when it was officially reestablished (on April 22, 1833) as a "trade town." The planned move to the new site was not well received by the inhabitants of the existing town of Cerneți, so it was not until 1836 that the new town began truly to take shape. A few years later, Turnu Severin became the county town of Mehedinți. The contribution of the DDSG to the town's economy increased in the 1850s, when it rented a piece of land on which it later built a modern shipyard.[62]

An even more central node was Giurgiu, which was restored to Wallachia in 1829 under the terms of the Treaty of Adrianople. Serving as the "port of Bucharest" and handling a considerable quantity of river traffic, Giurgiu enjoyed the special attention of travelers on the Danube. Some were lured by the fame of Bucharest, thinking it a pity to pass so close to the enchanting capital of Wallachia and not visit it.[63] Others were content to take advantage of the longer-than-usual stop to explore Giurgiu itself.

The Austrian Ida Laura Pfeiffer (1797–1858) became one of the most well-known European women travelers of the mid-nineteenth century after the publication of her interesting accounts of her journeys, which included a passage along the Lower Danube on the occasion of her trip to Jerusalem via Istanbul. On March 29, 1842, Pfeiffer embarked on the *Zrinyi* at Schela Cladovei, and two days later, in the morning of March 31, the steamer put down anchor at Giurgiu. As the operations of taking on coal and unloading eight carts and some tens of tons of goods lasted around eight hours, the passengers took advantage of the break to visit the town, which, for Pfeiffer, recalled the ugly and dirty settlements of Galicia: "The streets and squares are full of pits and holes; the houses are built without the slightest regard to taste or symmetry, one perhaps projecting halfway across the street, while its neighbor falls quite into the background."[64] Other travelers, too, left similar descriptions of Giurgiu, a town with "dirty narrow streets, and houses built of mud,"[65] in which the only "tourist attractions" of much interest seem to have been the clock tower and the ruins of the old citadel. However, the most important place in Giurgiu was the inn, the starting point for a trip to Bucharest. The journey to the capital familiarized the traveler with another memorable characteristic of travel through the Romanian lands: the cart, always part of a primitive and unholy trinity cursed by those who had to endure a boneshaking gallop across the Danubian Plain: road infrastructure, vehicles, and coachmen. On the way to Bucharest, the stopping place— another important topos not only for the sociability of the road but also for the formation of impressions of the country and its people—frequently features in travelers' accounts.

From April 1852, a diligence service concessioned to the entrepreneur Ştefan Burchi began to operate on the Bucharest–Giurgiu route. The conditions of the concession reflect the first attempts to modernize public transport in Wallachia: the coach must be fitted with "English springs, just as is customary in the other parts of Europe," with seating for ten inside and one beside the coachman. It was to circulate throughout the period in which navigation was open. "The coach will be harnessed with eight or ten post horses, according to the difficulty of the road, the horses being changed, according to need, at each post station, for a faster journey." The travelers, money bags, and luggage were entrusted to the care of "a coachman worthy and with good reference." There were two departures a week from Bucharest, on Wednesdays

---

64   Ida Pfeiffer, *Visit to the Holy Land, Egypt and Italy*, trans. H. W. Dulcken (London, 1852), 30–33.
65   Edmund Spencer, *Travels in Circassia, Krim Tartary, etc.*, vol. 1 (London, 1837), 76–77.

and Sundays, so that passengers could be taken to Giurgiu in time for steamers going downstream (Istanbul) or upstream (Vienna). For customers who wished to travel alone, the concessionary also offered additional options on request at double the price: coupés with two seats inside and gigs with four seats.[66]

Across the river from Giurgiu, the citadel of Rustchuk (Ruse) was, as already mentioned, a key point on the map of connections between the Balkan, Danubian, and north Danubian routes. Dimitrie Bolintineanu (1819–1872) left us a memorable account of a journey along the roads of the Balkans. His description underlines the fact that travel through Bulgaria remained difficult and unsafe, permitting a new comparison with the advantages of the river. Banished from Wallachia in the aftermath of the Revolution of 1848, Bolintineanu reached Ruse in November 1851, after traveling from Kladovo on the "Turkey line" steamer. "From Paris to Rustchuk!" the writer exclaimed irritably. "What difference and what change for the worse!" Bolintineanu hoped to be able to see his sister Caterina again, but after waiting in vain for a month, he decided to leave for Istanbul in the company of a German doctor whom he had met on board ship. The travelers hired a two-horse araba to go via Shumen and Varna, as navigation on the Danube was closed for the winter. Warned by an Ottoman pasha that the road was dangerous, they were soon convinced that the journey was more difficult than they had expected: "A road of thirty-one hours on foot, we covered in eight days." One of their most vivid memories was of "a winter's night, in the forest; the earth covered with snow; a large fire, beside which four people huddled as best they could, to escape the frost and the wolves; two horses constantly starting, snorting and trying to break their tethers; and some distance from the fire, several wolves gnashing their teeth and looking at the people and the fire." Close to Shumen, they had a confrontation with a group of robbers. Almost everywhere along the road, eating and sleeping conditions were very poor. In one Turkish khan, "we were given a room that lacked even the furniture customary in these rooms, the mat. [...] But the room had a brick stove, and this was a treasure for us." After spending ten days at Shumen, the travelers headed for

---

66 Constantin N. Minescu, *Istoria poștelor române. Originea, desvoltarea și legislațiunea lor* (București, 1916), 183.

Varna, facing similar adventures, with road accidents and more attempts at robbery along the way. At Varna, Bolintineanu and the German doctor stopped at a Turkish khan, before continuing on their way to Istanbul on board a steamboat.[67]

THE BLACK SEA LINE

At the Maritime Danube ports Brăila and Galați, passengers coming from Istanbul had to submit to a sanitary regime, which will be discussed in more detail in Chapter 5. The period of sanitary arrest varied according to the epidemiological situation in the region. Those travelers who were continuing their journey upstream were transferred in the lazaretto zone from the maritime to the river steamboat.

Galați was an important hub on the interimperial routes of southeastern Europe. Depending on the schedule and the weather conditions, travelers for the Ottoman Empire or the Ukrainian provinces were obliged to wait here for several days, which gave them time to visit the town. As already mentioned, the DDSG and Austrian Lloyd ships connected with those of the Russian steamboat company, which, from 1846, had sailings three times monthly between Galați and Odesa.

The development of foreign trade through the ports of the Maritime Danube encouraged systematic investment in improving the road infrastructure of the Principalities. This is more visible in the case of Moldavia, where Galați was the only commercial port connected to the great international markets. Maintenance works on the roads had begun during the period of Russian occupation (1828–1834) and continued under the rule of Mihail Sturdza. The Iași–Tecuci–Galați road, vital for the export of agricultural produce through the port of Galați, was modernized. A connection to Bukovina and the routes through the Habsburg Empire and the German space was provided by the Iași–Botoșani–Dorohoi–Herța–Mamornița–Chernivtsi road. Repair work was carried out on the Iași–Târgu Frumos–Roman–Târgu Ocna, Botoșani–Târgu Neamț, Tecuci–Focșani, and Bârlad–Galați roads, with the result that Moldavia was crossed by a network of what were

---

67 Dimitrie Bolintineanu, "Călătorii pe Dunăre și în Bulgaria," in *Călătorii*, vol. 1, ed. Ion Roman (București, 1968), 53–102.

considered relatively good public roads. In 1853, the Principality had 400 km of paved roads,[68] the "motorways" of the modern age.

Passenger transport services were also modernized. A diligence "after the European style" circulated daily between Galați and Brăila, proving useful to merchants who had business interests in both ports. In 1851, Neculai and Teodor Ghica obtained the privilege of running a diligence service on the Mihăileni–Iași–Galați route, which connected the Germanic space, via Galați, to the Orient.[69]

To return to the steamboat services to and from the Ottoman capital, the journey from Galați to Istanbul took around two and a half days, but could be delayed by the depth of the Sulina bar or by the storms that were frequently encountered on the Black Sea. In the 1850s, the Austrian Lloyd company's steamers offered weekly sailings on this route. They stopped at Tulcea and Varna before continuing toward the Bosphorus. In addition to the *Ferdinand I*, other vessels employed on this route included the *Ville de Vienne*, the *Schild*, the *Seri-Pervas*, the *Persia*, and the *Kollowrat*. Around 1845, the Austrian Lloyd had twenty-five steamers, with a total tonnage of 10,600 and a motor force of 3,310 hp.[70]

The most well-known vessel on this line was the *Ferdinand I*. Built in Trieste by the engineer Vincenzo Polli, it was launched for trials in February 1836 and entered commercial circulation a month later. It was around 44 m in length and 9.6 m in beam, with a draft of 2.3 m. It had a displacement of 308 tons and could carry a load of 141 tons. It was propelled by two paddle wheels using two horizontal motors produced by Boulton & Watt, with a combined power of 100 hp. The two boilers consumed around a ton of coal per day. It had a cruising speed of 9 knots (around 16 km/h) at sea and, in favorable conditions, could also make use of sails for propulsion. It was provided with sixteen first-class cabins and fifty places in the second-class common lounge, while third-class passengers traveled on the deck. It was captained first by the Englishman John Thomas Everson and later by the Italian Francesco Malombra.[71]

68 Paul Păltănea, *Istoria orașului Galați de la origini până la 1918*, 2nd ed., ed. Eugen-Dan Drăgoi, vol. 1 (Galați, 2008), 333–334; details on the roads of Moldavia in V. Popovici, C. C. Anghelescu, and L. Boicu, *Dezvoltarea economiei Moldovei între anii 1848 si 1864: contribuții* (București, 1963), 430–442.

69 Păltănea, *Istoria*, 2, 41–42.

70 *The Overland Mail and the Austrian Lloyd's* (London, 1847), 11.

71 Anastas Angelo, "Parakhodŭt 'Ferdinand I' po liniyata Konstantinopol—Varna—Sulina—Galats," http://morskivestnik.com/compass/news/2018/072018/images/ssFerdinand_08072018.pdf (accessed August 28, 2023).

Travelers recorded a great variety of opinions about the *Ferdinand I.* According to the British diplomat Charles William Vane, marquess of Londonderry (1778–1854), who sailed on board it in 1840, the steamer was neither large enough nor in any way suitable for carrying passengers on the Black Sea.[72] Octavian Blewitt, on the other hand, considered it "one of the best vessels of the Austrian company,"[73] while Ida Laura Pfeiffer writes, "Though not a large boat, the Ferdinand is comfortable and well built. Even the second-class cabin is neatly arranged, and a pretty stove diffused a warmth which was peculiarly grateful to us all."[74] I shall return to the question of traveling conditions in the next chapter.

For a number of years, the DDSG tried to make the Vienna–Istanbul route more efficient by carrying passengers overland between Cernavodă and Constanţa to bypass the region of the Danube Mouths. It was this route that the well-known Danish writer Hans Christian Andersen (1805–1875) took on his return from a journey in the Orient. On disembarking at Constanţa, Andersen and his fellow travelers could admire the small town, where "miserable, half-fallen-down houses formed the main street."

They spent the night in the agency's inn and then set out the next day on an interesting journey across Dobrogea. After stopping where a "little *khan* erected for us stood very invitingly on the way," they continued to the Danube, where the river steamboat *Argo* was waiting for them.[75]

Gheorghe Bibescu (1804–1873), prince of Wallachia in 1843–1848, also took the Dobrogea route when he traveled to Istanbul for his investiture at the Sublime Porte. The account by Simeon Marcovici (1802–1877),[76] who accompanied the princely suite as "press correspondent," is full of precise time references, the sign not only of a new relationship with the clock but also of an age in which travel had become more predictable. The steamer *Argo* left the port of Giurgiu on August 17, 1843, at 3 a.m. and arrived at Cernavodă the next day at 2.30 p.m. The travelers spent the night on board and then "the following

---

72 Charles William Vane, Marquess of Londonderry, *A Steam Voyage to Constantinople by the Danube and Rhine in 1840–41 and to Portugal, Spain etc. in 1839*, vol. 1 (London, 1842), 353.

73 Peregrine, "The Danube," 1, 560.

74 Pfeiffer, *Visit*, 36.

75 Hans Christian Andersen, *A Poet's Bazaar*, trans. Charles Beckwith Lohmeyer, vol. 3 (London, 1846), 99–115 (100, 113 for the quotes).

76 On Marcovici, see Alex R. Tipei, "Audience Matters: 'Civilization-Speak', Educational Discourses, and Balkan Nationalism, 1800–1840," *European History Quarterly* 48, no. 4 (2018), 658–685.

day, 19 August, at 7 hours of the morning, we began the journey on dry land to Chiustende [Constanţa]." The group arrived at the port at 3 p.m. and then

> at 6 ¼ hours after noon, we set out for Constantinople on the beauti-
> ful and elegant steamboat *Seri-Pervaz* ("Carry us fast"), which has the
> power of 160 horses. On 20 August, at 2 hours after noon we all arrived
> in good health and with the most pleasing weather at the mouth of the
> Bosphorus, or the channel of the Black Sea, and after an hour floating in
> this channel, we disembarked at the house of the Grand Logofăt Nicolae
> Aristarhi, the agent of Wallachia at the Sublime Porte.

Prince Bibescu returned to his home country by the same route, "weighed down with honors and precious gifts," arriving at Giurgiu at 10 a.m. on October 12, 1843.[77]

Istanbul was a great regional center of transport, with sailings to some of the greatest ports of the Black Sea (Sinope, Samsun, Trabzon) and the Eastern Mediterranean. It was thus possible to continue one's journey to Izmir or Alexandria, Athens or Malta, Trieste, Venice, or Marseille. According to Austrian Lloyd, in 1846 the company's steamers made thirty-six sailings between Istanbul and Galaţi/Brăila (eighteen in each direction), carrying 5,992 passengers, 4,430,073 florins in secure bags, 18,635 letters, 16,291 parcels, and 536 packets. The figures bear witness to the commercial importance of the route. In total, 326 travelers embarked that year at Brăila and 1,882 at Galaţi: in other words, an average of around 120 passengers on every sailing from the Maritime Danube ports to Istanbul (not counting those who embarked at Tulcea or Varna on the way). In the same year, 1,744 passengers disembarked at Galaţi and 300 at Brăila.[78]

A FORMIDABLE COMPETITOR: THE RAILWAYS

Trains appeared in the Habsburg monarchy not long after steamboats. In 1838, after the inauguration of the first section of railway in the empire (between Vienna and Wagram, part of the line connecting the imperial

---

77 Nicolae Isar, *Sub semnul romantismului: de la domnitorul Gheorghe Bibescu la scriitorul Simeon Marco-vici* (Bucureşti, 2003), http://ebooks.unibuc.ro/istorie/isar/index.htm (accessed August 28, 2023). References to other travelers through Dobrogea in Constantin Cioroiu, *Călători la Pontul Euxin* (Bucureşti, 1984), 173–176.

78 *The Overland Mail*, 8–10.

capital to Cracow), preparations began for the construction of a line between Vienna and Győr (Raab) with a branch line to Bratislava. The project advanced slowly, but the section as far as Bruck an der Leitha was completed in 1846. The same year saw the inauguration of the line between Budapest and Vác: the 33.6 km, with two intermediate stations, could be covered in just fifty-nine minutes. By the early 1850s, it was possible to travel by rail between Vienna and Budapest.[79] The train thus now offered the fastest and most comfortable connection between the great urban centers of the empire. The steamboat continued to attract travelers, both for the picturesque views it offered and for the more intense sociability on board. However, the changes in the structure of regional mobility forced the DDSG to adapt its business model, and it invested more and more in the goods transport side of its activity.

In the Banat, Oravița was linked to the Danube port of Baziaș by a 62.5 km railway, inaugurated in November 1856. Its role was to provide access to the river for the mining area around Anina, a vital project to supply coal for the industrial revolution in the Habsburg lands. The works were carried out by the company STEG (Staats Eisenbahn Gessellschaft), which had obtained a number of railway concessions in the empire. From 1858, after the inauguration of further lines on various intermediary segments, it was possible to travel by train between Vienna and Baziaș.[80] That same year, for example, according to an announcement in the Wallachian press, "all steamboats leaving Giurgiu have a connection at Baziaș with the train going to Pest, Vienna, Prague, Dresden, Leipzig, and Paris."[81] Another two decades were to pass before the railway reached Orșova, through the Cerna valley. Then, in 1879, a junction was made at Vârciorova between the railway networks of Austria–Hungary and of Romania to serve the mountain region of the Banat.[82]

With a growing range of options becoming available, travelers chose between the various means of transport according to timetables, budget, or convenience. That not everyone was enthusiastic about the qualities of the new means of transport is clear from the ironic account by the author Nicolae Filimon (1819–1865). "Somewhat disgusted" by his experience of

---

79 Botez et al., *Epopeea*, 43–44.
80 Ibid., 45–47.
81 BAPR, vol. 2, *1851–1858*, part 2, ed. Ioan Lupu et al. (București, 1971), 907 (*Anunțătorul român* 5, 1858).
82 Botez et al., *Epopeea*, 48.

the steamer *Archduke Albert*, on which he had traveled from Giurgiu to Hungary, Filimon continued his journey to Vienna by rail. "I had until then traveled only in Brașov wagons and carriages with springs; I thus imagined that the iron road would be incomparable in speed. This idea I had from travelers from our country, who in their eagerness to utter paradoxes had assured me that the railway wagons ran much faster than the wind and somewhat slower than the mind."[83] Comparing the speed of the Austrian trains and the "Romanian post," Filimon concludes: "The Austrian railway is three times faster than the post in our country."[84] This was progress, to be sure, even if it did not quite live up to his expectations.

In the region of the Lower Danube too, the river was integrated into a more and more complex transport network. The plan of some British entrepreneurs to build a canal between Constanța and Cernavodă was rejected. However, the project of making a rail connection across Dobrogea was put into practice starting in 1857. The railway, about 65 km long, was inaugurated in October 1860. It was now easy for travelers going between Central Europe and Istanbul to bypass the Danube mouths. This came as a blow to the transport hubs of the Maritime Danube, especially Galați, which was trying to maintain its position in the European transport network. Six years later, the same entrepreneurs completed the railway line between Ruse and Varna, thus enabling travelers in a hurry to reach Istanbul to bypass Dobrogea altogether. Significant investments were also made in Romania, where the railway between Bucharest and Giurgiu became operational in November 1869.[85] Two trains ran each way daily, covering the distance of 70 km in an hour and three-quarters. The Danube remained the central axis of a more and more complex transport infrastructure, but the railways were creating new options for fast, cheap, and safe travel.

New railway lines were also constructed in Bukovina. Czernowitz (Chernivtsi) was connected to Vienna from 1866, and the line was soon extended into Romanian Moldavia. Railways were built on a massive scale in Transylvania in the 1860s and in Romania in the following decade.[86]

---

83 Nicolae Filimon, *Escursiuni în Germania meridională. Nuvele*, ed. Paul Cornea (București, 1984), 59. The option of traveling faster than the wind or faster than the mind (literally "than thought") is a commonplace of Romanian folktales.
84 Filimon, *Escursiuni în Germania meridională*, 60.
85 Botez et al., *Epopeea*, 68–74.
86 Ibid., 51–65.

The crowning moment of this engineering and entrepreneurial frenzy came in June 1883 when the *Orient Express* first departed from Paris along the route Strasbourg–Stuttgart–Munich–Vienna–Budapest–Vârciorova–Bucharest–Giurgiu–Smârda. After being ferried across the Danube, passengers continued by train from Ruse to Varna and then by steamer to Istanbul. From West to East, via Smârda, in less than eighty-four hours (three and a half days) was, it must be acknowledged, an impressive achievement, demonstrating not only the excellent international coordination of rail companies but also the fact that the Danube had now ceased to be the central axis of West–East travel.

CONCLUSIONS

The success of the steamboats was due to the combination of three advantages they had in comparison with travel overland in animal-drawn vehicles: speed, comfort, and relatively accessible price. For the route between Vienna and Istanbul, as for other routes in peripheral areas, there was also the matter of safety on a journey through territories that were still little known to Western travelers. The commodification of travel meant the inclusion of certain guarantees of safety and comfort in the price of the ticket. The average speed of the vessels was not high, but the fact that they were able to sail from dawn till dusk (and at sea, even during the night) meant that they could cover some 200 km per day. The mechanization of traction and the relatively decent level of comfort offered on board meant that no great consumption of human energy was required, and the travelers' bodies were no longer subjected to the physical and mental exhaustion caused by constant jolting along the public highways. Temporal and financial predictability were part of this form of comfort and of a new management of time, which allowed the making of clear travel plans, starting from a generally respected timetable and costs known in advance. The price of tickets, it should be emphasized, was accessible, opening up the routes of mobility to representatives of all social categories. The inclusion of safety in the price was an important aspect, given the fears associated with the dangers that might be encountered along the roads of the Ottoman Empire. With major sources of physical and mental stress thus eliminated, a steamboat voyage on the Danube became an interesting social experience and an aesthetic spectacle in which passengers could

enjoy the cosmopolitan mix of fellow travelers and could admire at their leisure the landscape that opened up before them.

Opinions about all these relative advantages varied greatly, depending on
the travelers' expectations. For some, like the British travel writer Edmund
Spencer (?–?), the voyage was a pleasant one:

> I have seldom performed a tour which afforded me more real pleasure,
> nor one that offered scenes of such varied interest, whether we regard
> the beauty of the scenery, the striking diversity of features exhibited by
> the different provinces, together with the primitive state of the inhabit-
> ants; the whole passing in review as if in a panorama. Nor must I forget
> to mention, that the whole expense attending the voyage amounted to no
> more than about eleven pounds.[87]

Other passengers were critical of failings that could be blamed on the
DDSG's management and other travel arrangements. Others still, for various reasons, preferred to stick as long as possible to the overland routes,
even if these continued to have many of the disadvantages mentioned in this
chapter. James Baillie Fraser (1783–1856), a Scottish travel writer and art-
ist with an excellent knowledge of the Middle East, traveled from Vienna
to Istanbul as part of a longer journey from London to Tehran, accompanying three Persian princes on their journey back to their home country. Arriving in Vienna in September 1836, the princes decided that the
steamer would not offer sufficient comfort and safety. The group thus tra-
versed the Habsburg Empire by carriage, passing through Budapest, Szeged,
Timișoara, and Sibiu, only to be held up in Bucharest because of the plague
epidemic that was haunting the Balkans. Fraser's account underlines many
of the inconveniences of the overland journey, resulting not only from the
poor quality of roads and vehicles but also from the accommodation infrastructure available along their route. Finally, the three princes and their
entourage embarked at Galați on the *Ferdinand I*, which took them to
Istanbul.[88]

---

87 Spencer, *Travels*, 90–91.
88 James Baillie Fraser, *Narrative of the Residence of the Persian Princes in London, in 1835 and 1836*, vol. 1,
    2nd ed. (London, 1838), 86–217.

Other travelers too opted for overland routes on their journeys from the West to the Orient. However, a quantitative analysis of travel accounts in the series "Călători străini despre țările române" ("Foreign travelers about the Romanian lands") published by the "Nicolae Iorga" Institute of History of the Romanian Academy shows that of the travelers who passed through the Romanian lands in 1836–1851 (included in volumes 3–5 of the new series, covering the period after the launch of the steamboat service between Vienna and Istanbul), around two-thirds did so on board the Austrian steamers.

Despite the multiple problems about which travelers complained, to which I shall return in the following chapters, the Danube route was a great success for the DDSG, a shipping company that facilitated not only the movement of travelers but also the transfer of up-to-date technology in southeastern Europe. The DDSG made maximum use of the natural highway of the Danube. Not having to invest in the maintenance of the channel (for which the imperial government took responsibility), the company developed rapidly, investing in coal mines to ensure the supply of fuel for its ships, and in shipyards for the construction of new passenger and goods vessels. In the early years, the same steamers carried both passengers and goods, but toward the middle of the nineteenth century, an increasingly clear specialization may be observed, as the company invested in improving the quality of passenger transport precisely by separating the two types of business.

The DDSG steamers opened up for the Habsburg Empire a new route of economic and political expansion the Austrians had long dreamed of. To paraphrase Freda Harcourt's observations about another shipping company, the steamers were the messengers and defenders of Austrian hydro-imperialism in the regions of the Lower Danube and the basin of the Black Sea and Eastern Mediterranean.[89] The Vienna government's interest in transforming the Danube into an international transport infrastructure was great. Borrowing from the analysis of Joanna Guldi, who has demonstrated how the development of transport infrastructure and the creation of an efficient system of public roads contributed to the consolidation of Great Britain in the eighteenth century,[90] we may consider that the Austrian "infrastructure

---

89   Freda Harcourt, *Flagships of Imperialism: The P&O Company and the Politics of Empire from Its Origins to 1867* (Manchester, 2006).
90   Joanna Guldi, *Roads to Power: Britain Invents the Infrastructure State* (Cambridge, MA, 2012).

state" took shape in the following century, with the great river as the spinal column of the Danubian monarchy.[91]

The DDSG invested in greater efficiency in all the key aspects of its business model: speed, comfort, accessibility, and safety. In two decades, it managed to halve traveling times on the Vienna–Istanbul route: the faster steamers of the 1850s could cover the distance in around seven days. Thanks to investment in modern vessels, together with arrangements to speed up operations in the ports and improved punctuality in connections, the comfort of steamboat travel was considerably increased.

By around the middle of the nineteenth century, however, the DDSG had done more or less everything that was economically and technically possible to improve conditions of travel and had effectively reached the limits of the technology on which it relied. Railways now connected the main cities of Central Europe, and investors were coming up with new projects to bring trains to the southeast European periphery. Railways were more advantageous from the point of view of the four aspects mentioned above: speed, comfort, accessibility, and safety. In addition, they came with many other advantages, including the fact that they were more profitable. Thus, the steamboat saw a gradual decline, on the Danube and elsewhere, caused by the same factor that had ensured the success of the DDSG and other similar companies: technological innovation, which increased the speed of travel even more, ensuring greater comfort, accessibility, and safety in the transport of people and goods. The railways gradually came to be the leading means of inland passenger transport, while steamboats acquired a niche for themselves in medium- and long-distance coastal trade, to which they were better adapted than trains.

---

91 Robert Mevissen, *Constructing the Danube Monarchy: Habsburg State-Building in the Nineteenth Century,* PhD thesis (Georgetown University, 2017). Also Mevissen, "Meandering Circumstances, Fluid Associations: Shaping Riverine Transformations in the Late Habsburg Monarchy," *Austrian History Yearbook* 49 (2018), 23–40.

CHAPTER 2

A Floating City

SNORES AND IDENTITIES

The sound of snoring is one of those sounds that are unanimously detested.[1] I do not believe there is any culture in which snoring signifies and generates anything other than trouble, generally domestic, as the family is forced to endure the snorer's "performance." The modern world with its new means of transport turned snoring into an issue of public interest, even one of sociocultural identity. No one has captured this better than the Moldavian poet Vasile Alecsandri, describing a memorable experience at nightfall in the cabin of the steamboat *Széchenyi*, on which he embarked at Schela Cladovei in November 1851, on his way back from Paris to Iași:

> Then, one by one all [the passengers] stretched out on benches, on chairs, and on tables; the oil lamp went out as never before and in the heart of the darkness there rose a huge snore, composed of a variety of snores, baptized and unbaptized alike. The room resounded like ten stoves roaring and ten mills grinding! Just try sleeping, if you can, in the midst of such harmonies. As for me, after I had made every attempt to fall asleep, after I had rolled time and time again now to one side, now to the other, after I had tried to count to a thousand, in vain! [...] I lit my cigarette and set about guessing the nationalities of the various snores that were giving me such a melodious serenade. A new and interesting study, which

---

[1] This chapter is expanded and adapted from my recently published article: Ardeleanu, "'Steamboat Sociality' along the Danube and the Black Sea (mid-1830s–mid-1850s)," *Journal of Transport History* 41, no. 2 (2020), 208–228.

I recommend to all those unfortunate travelers who are condemned by circumstances to whole nights of sleeplessness.

So, at my feet, there sighed a musical snore resembling a scale from do to mi, which sometimes sounded as if it were about to begin an aria from *The Barber of Seville*, but that deceptive opening immediately ended with a sorrowful note from *Lucia*. Who could that dilettante snorer be? Undoubtedly an Italian! And indeed, it was a young man going to Bucharest. [...] Beside the door, there piped a snore as sharp and thin as the hiss of a snake, and for all that, it inspired a sort of merciful compassion. It irritated you and at the same time it stole your attention; it attracted you and at the same time it strained your nerves; the man who snored like that had to be dangerous; a monster with two faces, with sweet lips and envenomed teeth, with mild eyes and a duplicitous heart! What could that stranger be? [...] a highway robber or a spy? What was his nationality, his trade?[2]

The snoring in the communal cabin of the steamboat is a common theme in the literature of travel on the Danube. Three brief references are, I hope, equally vivid. In the cabin of the steamer *Franz I* on a torrid July evening in 1837, the British doctor William Fullerton Cumming (?–?) found "the atmosphere positively pestilential" and was prevented from sleeping by the "strange, unearthly sounds" made by a fellow traveler lying on the floor who "snored like a rhinoceros." The situation was "laughable for a time," but soon became serious enough for another passenger to intervene and waken the offending snorer.[3] Xavier Marmier (1808–1892) similarly recalled in 1846, after a journey between Budapest and Zemun, the "frightful cacophony" of forty-two noses—German, Hungarian, and Slav—delivering "the most fantastic concert on all the notes of the scale, from the falsetto of the choir boy to the bass of Lablache." One nose stood out in particular, "a sort of pimply trumpet, planted on the face of a timber-merchant who seemed to sound the charge and beat the rhythm." The other noses followed as best as they could, some like trombones, others like violins, while one intervened at intervals

---

2   Vasile Alecsandri, "Înecarea vaporului Seceni pe Dunărea" ("The Drowning of the Steamboat *Széchenyi*"), in *Proză. Povestiri. Amintiri romantice*, ed. Alexandru Marcu (Craiova, 1939), 77–78.

3   William Fullerton Cumming, *Notes of a Wanderer, in Search of Health, through Italy, Egypt, Greece, Turkey, up the Danube and down the Rhine* (Edinburgh, 1839), 238.

with "a prolonged and sonorous vibration, like the sound of a tom-tom." Having had enough of the concert, Marmier took refuge on the deck, preferring to sleep in the open air.[4] So did the American doctor James O. Noyes (1829–1872), traveling on the Danube in 1854:

> But more amusing was it to listen to the cacophonia of fifty noses of all sizes, nationalities, shades of color and varieties of tone. Snoring seemed contagious; I could only liken the saloon to a vast æolian harp, or to an orchestra of dead men, playing funeral dirges upon the harshest wind instruments. I fancied that I could trace in their monotonous discords the influence of wine, and love and sorrow, of nightmare visions from distended stomachs, and of beautiful dreams weaving their golden threads in the gossamer tissues of the brain. It was in vain that I covered my head and stopped my ears. More than once during the long watches, I left the hot and mephitic cabin to enjoy on deck the sweet influences of night and of the stars.[5]

My aim in this chapter is to present other such unconventional encounters in the Austrian steamboat cabin, which we may think of as a mobile space with an intense social life. The idea is far from new in literature or in cinematography, to mention two media in which the theme of sociality on board steamers has been put to good use. In the 1860s, Jules Verne described his voyage on the *Great Eastern*, the gigantic steamship on which he traveled to America. The vessel was a veritable "masterpiece of naval construction," but more than that, it was a "floating city" with an intense social life: "If the *Great Eastern* is not merely a nautical engine, but rather a microcosm, and carries a small world within it, an observer will not be astonished to meet here, as on a larger theatre, all the instincts, follies, and passions of human nature."[6] More recently, James Cameron, in one of the most well-known and profitable films in history, has depicted the social life on board the *Titanic*. Quite apart from questions of artistic value, the film clearly offers an excellent picture of everyday life and the various forms of social interaction on

---

4   Xavier Marmier, *Du Rhin au Nil: Tyrol—Hongrie—Provinces Danubiennes—Syrie—Palestine—Egypte. Souvenirs de voyages*, vol. 1 (Bruxelles, 1852), 185–186.
5   James O. Noyes, *Roumania: Border Land of the Christian and the Turk* (New York, 1857), 67.
6   Jules Verne, *A Floating City*, translation from French (London, 1918), 11.

board the famous British ship. Even if they cannot be compared for scale with the *Great Eastern* or the *Titanic*, the Austrian steamers that sailed on the Danube and the Black Sea are equally fascinating historical settings for a study of the sociality generated by the means of transport characteristic of the modern world.

THE STEAMBOAT AS THEATER OF GLOBAL HISTORY

Through their utilization for commercial purposes, steamboats were fundamental to the transport revolution, which, continuing with the introduction of railways, profoundly changed the world. Steam-powered vessels were the first agents of modernization,[7] accelerating regional and global mobility due to their ability to transport passengers and goods relatively fast, safely, and cheaply. At the same time, they facilitated access to new types of social contact for the growing numbers of customers who traveled on them, as they crossed lands and seas for business or for pleasure.[8]

In this chapter, the Austrian steamers will serve as arenas of global history through whose intermediary I shall explore the social dimension of travel. In the first place, I shall look in detail at the forms of sociality created by the introduction of steam navigation on the Danube. Sailing along a river—which was at the same time both border and connecting highway between the Austrian, Ottoman, and Russian Empires—the steamers themselves constituted a floating world, with an intense social life. As sociality is an intrinsic component of modern transport, the chapter will foreground aspects of daily life relating to the early phases of the industrialization and commodification of travel,[9] with an emphasis on the social experiences travelers encountered as they sailed between empires and civilizations.

For Michel Foucault,[10] ships, as floating spaces with their own identity, are classic examples of heterotopias. More recently, historians have

---

7   John Armstrong and David M. Williams, "The Steamship as an Agent of Modernisation, 1812–1840," *International Journal of Maritime History* 19, no. 1 (2007), 145–160.

8   Armstrong and Williams, "The Steamboat and Popular Tourism," *Journal of Transport History* 26, no. 1 (2005), 61–77; Williams and Armstrong, "'One of the Noblest Inventions of the Age': British Steamboat Numbers, Diffusion, Services and Public Reception, 1812–c. 1823," *Journal of Transport History* 35, no. 1 (2014), 18–34.

9   Mackintosh, "Ticketed Through," 61–89.

10 Foucault, "Of Other Spaces," 27.

also made use of the concept of the contact zone to describe the ship as a meeting place of diverse individuals and sociocultural environments. Thus, "the passage itself" has been examined in an article by Michael David Offermann and Roland Wenzlhuemer, for whom the steamship is "a historical arena, a zone of cultural contact and interaction," with all "the transcultural phenomena arising as a consequence."[11] Through the spaces, moments, or journeys they produce, ships have also been studied by historians from the point of view of mobility. In their introduction to a recent volume, Anyaa Anim-Addo, William Hasty, and Kimberley Peters draw a distinction between "the mobilities of ships and shipped mobilities," insisting on the vessel's function as a "global connector."[12] Martin Dusinberre and Roland Wenzlhuemer add that ships have been studied only as mobile objects that connect places, and not as "historically relevant arenas in themselves." We may thus note a growing academic interest in the realities and interactions on board ships, seen as "mobile spaces," as laboratories in which and through which historians can follow processes of historical transformation. Connections give rise to new forms of intermediation, and the state of being in movement or in transit brings with it special experiences,[13] aspects that are generally invisible in historiography. Another relatively recent study has analyzed the evolution of the spatial and temporal experiences of Chinese passengers on board steamboats: travelers and settings that transported new ideas about progress, international law, race, and civilization.[14]

Despite the growing interest shown by researchers applying modern concepts and approaches, the study of the steamboat as a mobile space is still in its early stages, at least in comparison with the attention given to other means of transport. Studies in the history of railway and automobile transport

---

11   Michael David Offermann and Roland Wenzlhuemer, "Ship Newspapers and Passenger Life aboard Transoceanic Steamships in the Late Nineteenth Century," *Transcultural Studies* 3, no. 1 (2012), 89.

12   Anyaa Anim-Addo, William Hasty, and Kimberley Peters, "The Mobilities of Ships and Shipped Mobilities," *Mobilities* 9, no. 3 (2014), 337–349.

13   Martin Dusinberre and Roland Wenzlhuemer, "Editorial: Being in Transit: Ships and Global Incompatibilities," *Journal of Global History* 11, no. 2 (2016), 155–162. See also Wenzlhuemer, "The Ship, the Media, and the World: Conceptualizing Connections in Global History," *Journal of Global History* 11, no. 2 (2016), 163–186.

14   Jenny Huangfu Day, "From Fire-Wheel Boats to Cities on the Sea: Changing Perceptions of the Steamships in the Late Qing, 1830s–1900s," *Australasian Journal of Victorian Studies* 20, no. 1 (2015), 50–63.

(by such influential authors as Wolfgang Schivelbusch,[15] David Bissell,[16] Mimi Sheller, and John Urry[17]) have resulted, especially with regard to mobility, in new theoretical approaches that insist on the experiences of train and car passengers, experiences that oscillate between the public and the private, the known and the unknown, the national and the global.[18] I would also draw attention to Radu Mârza's recent book about Romanian perceptions of railway travel, which is all the more interesting inasmuch as it is a field that has not yet been much explored in Romanian historiography.[19]

The relative absence of studies of mobilities on water has been remarked on by historians. Few researchers have directed their attention to "the time of the passage itself, let alone to questions of sociocultural interaction on board."[20] And for those who have, the main focus of interest has been on the intercontinental voyages of the late nineteenth and early twentieth centuries.[21]

The aim of this chapter is thus to analyze some of the components of a special form of mobility on water and to examine in detail the manner in which steamboat journeys modeled sociality on board. I shall refer to a wide range of social interactions on the Austrian steamers that sailed between Vienna and Istanbul from the 1830s to 1860. Given the multitude of descriptions available, I shall make use of those written by travelers who covered several portions of the route between the two capitals and thus spent a number of days and nights on board the steamer as direct participants in its social life.

Three main factors transformed a voyage on the Danube and the Black Sea into a special experience in the middle decades of the nineteenth

---

15 Wolfgang Schivelbusch, *The Railway Journey: The Industrialization of Time and Space in the Nineteenth Century* (Oakland, 2014).

16 David Bissell, "Moving with Others: The Sociality of the Railway Journey," in Phillip Vannini (ed.), *The Cultures of Alternative Mobilities. Routes Less Travelled* (Farnham, 2009), 55–69.

17 Mimi Sheller and John Urry, "The City and the Car," *International Journal of Urban and Regional Research* 24, no. 44 (2000), 737–757; Sheller and Urry, "The New Mobilities Paradigm," *Environment and Planning A* 38, no. 2 (2006), 207–226.

18 Vincent Kaufmann, "On Transport History and Contemporary Social Theory," *Journal of Transport History* 28, no. 2 (2007), 302–306.

19 Radu Mârza, *Călători români privind pe fereastra trenului. O încercare de istorie culturală* (Iași, 2020).

20 Offermann and Wenzlhuemer, "Ship Newspapers," 89.

21 Mark Rennella and Whitney Walton, "Planned Serendipity: American Travelers and the Transatlantic Voyage in the Nineteenth and Twentieth Centuries," *Journal of Social History* 38, no. 2 (2004), 365–383; for an analysis of the historiography, see Offermann and Wenzlhuemer, "Ship Newspapers," 89–91.

century, making the journey by river and sea an interesting theme from
the point of view of the history of transport. First, steamboats provided a
relatively fast, safe, and comfortable connection between Western Europe
and the Levant and thus carried a wide variety of passengers. Second, compared with voyages on other great rivers of Europe and North America,
the Danube route was sufficiently long to allow passengers the leisure and
comfort to establish a wide diversity of social bonds. Third, by reason of
their construction, the Austrian steamers were characterized by open spaces
rather than private cabins, thus making socialization an integral part of the
voyage. But apart from these specific aspects, similar social encounters were
common to all the incipient phases of the commodification of travel on such
rivers as the Ohio, the Mississippi, and the Rhine.[22] In the following pages, I
shall describe the spaces and the moments of social encounter on board the
Austrian steamers.[23] I shall then turn to situations of intense socialization,
with details about aspects specific to river travel in comparison with maritime and intercontinental voyages and about the social relevance of travel
along borders of great strategic significance for the empires of Europe. I thus
hope to contribute to the idea that the steamers served as spaces of connection and intermediation as well as for examining in detail the experience of
being in transit on a mobile stage floating along a fluid space.

MOBILE SPACES OF SOCIALIZATION

By selling a wide range of tickets at prices accessible to all pockets, the DDSG
attracted a large clientele, comprising passengers of diverse ethnicities, reli-
gions, and social categories. The statistical details provided in Chapter 1
point to the differences, in both quantitative and qualitative terms, between
the Habsburg and Ottoman segments of the journey. The busiest route was
that between Vienna and Budapest, followed by the "Hungarian" Danube
from Budapest to the entrance of the Danube Gorge. At the beginning of the

---

22   Hagen Schulz-Forberg, "The Sorcerer's Apprentice: English Travellers and the Rhine in the Long Nine-
     teenth Century," *Journeys* 3, no. 2 (2002), 86–110; Thomas C. Buchanan, *Black Life on the Mississippi:
     Slaves, Free Blacks, and the Western Steamboat World* (Chapel Hill, 2004); Rita Kohn (ed.), *Full Steam
     Ahead: Reflections on the Impact of the First Steamboat on the Ohio River, 1811–2011* (Indianapolis, 2011);
     Douglas R. Burgess Jr., *Engines of Empire: Steamships and the Victorian Imagination* (Stanford, 2016).
23   Robert Burroughs, "Travel Writing and Rivers," in N. Das and T. Youngs (eds.), *The Cambridge History
     of Travel Writing* (Cambridge, 2019), 330–344.

1840s, the DDSG had twenty vessels operating on these internal routes, with most passengers (around 90 percent) traveling first or second class. Along the Lower Danube, between the Iron Gates and Galați, more than half traveled third class, that is, on the deck. The proportion of third-class passengers rose to 85 percent for the journey between Galați and Istanbul. The complete voyage from the West to the Orient was thus marked by a gradual drop in both the number and the status of the passengers, with more and more of them traveling on the deck. Passengers who embarked on the entire journey observed this change; there was no need for them to leave the steamer to witness the transition to a world that was different not only culturally but also socially. For the French architect Félix Pigeory (1806–1873), on the Middle Danube, "passengers, faces, and costumes of Europe" still predominated,[24] but beyond the Iron Gates, the ethnic and religious variety of the passengers became more and more striking, making the voyage an immersion in the "exotic" Orient, seen with all the clichés and prejudices that Edward W. Said refers to in his classic work.[25]

In the accounts of Western writers, the journey between East and West, to look at it now from the Ottoman end, seemed to be one into the interior, a form of mobility from the deck to the cabin and from the open air to the elegant saloon. The interior architecture of the steamers was designed to maximize their profitability, with the result that the accommodation options were less varied than on the great ocean-going liners of the same period.[26] Nevertheless, there were several levels of stratification and segregation, not only with differences deriving from price classifications but also with spaces adapted to gender and religious norms, so as to satisfy the expectations of a global clientele. This design contributed to increased sociability on board, turning the steamer into a space of connections and intermediations, not only between different places but also among the members of the cosmopolitan group who shared a common social experience in the course of their journey.

Private cabins were rare and often reserved by or for important guests. Opinions about their comfort differed considerably. After a dispute over

---

24 Félix Pigeory, *Les pèlerins d'Orient, Lettres artistiques et historiques sur un voyage dans les Provinces Danubiennes, la Turquie, la Syrie et la Palestine* (Paris, 1854), 46.

25 Edward W. Said, *Orientalism* (London, 1978).

26 Douglas Hart, "Sociability and 'Separate Spheres' on the North Atlantic: The Interior Architecture of British Atlantic Liners, 1840–1930," *Journal of Social History* 44, no. 1 (2010), 189–212.

accommodation, the captain of the steamer *Pannonia* gave up his own cabin to Frances Anne Vane, marchioness of Londonderry (1800–1865), traveling on the Danube in 1840 with her husband, a notable British diplomat. She vividly describes it as "a den, full of the dirty captain's dirtier things [...] more like a wooden box or dog-kennel than any thing else." Torrential rain flooded the cabin, leaving her with only the vain hope that at least the fleas might be drowned![27] On the other hand, the German countess Ida von Hahn-Hahn (1805–1880), traveling in 1843, was content with her cabin, which at least offered her a small intimate space and sheltered her from the crowd in the ladies' saloon.[28] For the young Wallachian boyar Constantin A. Rosetti (1816–1885), "The best voyage of all that have been discovered till today is the steam vessel, having also a cabin. You neither feel exhaustion nor waste time. It seems as if it is a sort of magic, you leave one room and enter another room, hundreds of hours away."[29] The cabin thus became an almost charmed space, a sort of magic carpet that enabled one to fly through space and time, without any of the weariness of overland travel.

Public spaces, however, predominated on board the Austrian steamers (Figures 4 and 5). The main site of social encounters was the saloon. This served a number of functions: lounge, dining room, and bedroom. Travelers gave considerable attention to the saloon. As described by Hans Christian Andersen, in the cabin of the *Ferdinand I*, on board which he sailed on the Black Sea in 1841: "Elastic divans, and convenient hammocks surrounded a large ornamental saloon with mirrors, pictures, and books; fresh Egyptian figs, plucked a week before, were set out on the table, with grapes from Smyrna, and wine from the far distant Gaul." On the Lower Danube, the river steamboat *Argo*, onto which Andersen and his traveling companions were transferred at Cernavodă, likewise had "a saloon with mirrors, books, maps, and elastic divans; the table was spread with steaming dishes, fruits, and wine."[30]

---

27   Vane, *A Steam Voyage*, 124–126; Frances Anne Vane, *Narrative of a Visit to the Courts of Vienna, Constantinople, Athens, Naples etc. by the Marchioness of Londonderry* (London, 1844), 77–78.

28   Ida von Hahn-Hahn, *Letters from the Orient: Or, Travels in Turkey, the Holy Land and Egypt*, trans. S. Phillips (London, 1845), 19–20.

29   C. A. Rosetti, "Note intime scrise zilnic (1844–1859)," in *Lui C. A. Rosetti: la o sută de ani dela nașterea sa* (București, 1916), 184.

30   Andersen, *A Poet's Bazaar*, 93, 115.

Figure 4 Dieudonne Lancelot, *Second-Class Passengers on a Danube Steamer* (1860).

Figure 5 Lancelot, *Deck of a Danube Steamer* (1860).

om Istanbul to Vienna in 1838. One of the interesting characters
on board was Nadir Achmet Bey, who would later be involved in all sorts
of political intrigues in the Ottoman Empire and was suspected by some
passengers of being a Russian spy. At dinner on deck, Nadir Bey impressed
Romer with his knowledge of English literature and gentlemanly manners.
Under the influence of the wine he had drunk, however, he went on to order
his servant to place his hunting falcon, of which he was very proud, on the
table. Some of the diners protested, and a quarrel ensued, which the captain tried to mediate without success. Visibly suffering from the effects of
the alcohol he had consumed, Nadir Bey fell asleep on the deck, tended by
his faithful servant, leaving the Western passengers shocked at his display of
degraded Oriental manners.[39]

Stephen Olin (1797–1851), Methodist pastor and president of Wesleyan
University, Connecticut, had among his traveling companions on his journey up the Danube a Turkish pasha, "a fine specimen of the semi-barbarous,
half-civilized class of Turks, which has sprung up under the reforms of the
late grand sultan," and a young Circassian woman, purchased at the slave
market of Istanbul to become the wife of the pasha of Vidin. The "aged and
respectable-appearing duenna" who had been entrusted with the transac-
tion, an Ethiopian eunuch, and a black slave woman from the bride's retinue
proved equally interesting examples for a study of Oriental manners.[40]

Five years later, in 1845, the Scottish writer Felicia Skene (1821–1899)
sailed up the Danube on her return home after six years in Greece. The young
woman observed attentively the diverse "members of our little community
on board," who "furnish us not only with amusement, but also with food
for serious reflection." A family of Turks who embarked at Ruse, consisting
of the head of the family with his wives and children, provided her with an
occasion to observe at close quarters the private customs of Muslims. The
"lord and master" was a high-ranking gentleman with a passion for astron-
omy and turned out also to be a skillful chess player, but his wives appeared
to Skene to be completely ignorant and uncivilized.[41]

---

39 Mrs. [Isabella Frances] Romer, *The Bird of Passage or, Flying Glimpses of Many Lands*, vol. 2 (London,
     1849), 138, 150–160.
40 Stephen Olin, *Travels in Egypt, Arabia Petræa, and the Holy Land*, vol. 2 (New York, 1843), 473–475.
41 Felicia Skene, *Wayfaring Sketches among the Greeks and Turks and on the Shores of the Danube* (London,
     1847), 242, 255–257, 299.

From the saloon to the deck, the steamer was something of an aquarium in which various social species, incompatible in other environments, lived together for the duration of the voyage, each in sight of the others. The saloon is presented as a space of civility, with Western etiquette and habits, while the deck is characterized by a sort of Oriental barbarism or ignorance. The steamer had its own social geography, and a walk on board was a veritable journey not only between cultures but also through contact zones that intermediated a great variety of social encounters. With the exception of the more accentuated cosmopolitanism of the Ottoman sections of the route, it was no different in this respect from the steamboats that circulated on other rivers in North America or Western Europe.[42]

AGENTS OF SOCIALIZATION ON A "FLOATING BABEL"

Members of the crew played a crucial role not only in the maneuvering of the vessel but also in the intermediation of social life on board. It was they who made connections among the passengers, according to a well-established mobile hierarchy,[43] and this social function of theirs was part of the commodified service purchased by customers. Crews were extremely diverse ethnically. Britons, Italians, and Dalmatians captained DDSG steamers and, from a social point of view, played host to the most distinguished travelers. The Scottish clergyman George Robert Gleig (1796–1888) recalled that the captain of the steamer on which he sailed between Budapest and Belgrade in 1837 was

> a native of Florence, a singularly handsome man; of very agreeable manners, and nowise disposed to underrate his own accomplishments. An excellent linguist, he could converse with facility in Italian, English, German, French, Spanish, and even Hungarian. I do not know what his merits might be as a seaman, but his attention to his passengers was throughout unremitting, and he earned for himself in consequence

42 Annalies Corbin, *The Material Culture of Steamboat Passengers: Archaeological Evidence from the Missouri River* (New York, 2002), 9–10; Louis C. Hunter, *Steamboats on the Western Rivers: An Economic and Technological History* (Newburyport, 2012).

43 John Urry, "Social Networks, Mobile Lives and Social Inequalities," *Journal of Transport Geography* 21 (2012), 24–30.

golden opinions everywhere, especially among the fairer portion of the creation.[44]

The crew included engineers, pilots, gendarmes, servants, and restaurant staff, all of whom had both nautical and social roles. The Russian diplomat and industrialist Anatol Demidov (1813–1870) writes: "For its part, the crew, made up of men of all sorts of nations, nonchalantly obeys the captain, whose commands can be transmitted to this floating Babel only by means of three or four different languages."[45] For de Perthes, "the real lord of the place was the purser, an elegant young man from Vienna, who spoke Italian well and French a little" and who managed to deal with passports and provisions and to attend to the passengers without neglecting his spaniel.[46]

The passengers themselves were important agents of socialization. On board each vessel there were individuals who drew the attention of the whole company. Such VIPs included princes, ambassadors, pashas, and high-ranking officers, but even more important were those travelers whose natural volubility immediately made them the center of attention. One such was a "Moldavian adventurer" whom Michael J. Quin's observed on his Danube voyage:

He whistled well, he sung well, and passed off every thing in a "devil-may-care" kind of way, which gained him admirers. [...] He had a commonplace-book in his bosom—for his pockets had all vanished—from which he occasionally read to his followers scraps of poetry of his own composition, or selected from the works of celebrated German writers. These readings he interspersed with comments often so droll, that he set the whole deck in a roar.[47]

Such communicative characters were crucial to the emergence of a feeling of community among the passengers and to social mobility between the various groups and spaces on board the steamer. Bolintineanu also

44 George Robert Gleig, *Germany, Bohemia, and Hungary Visited in 1837*, vol. 3 (London, 1839), 263–264.
45 Anatole de Démidoff, *Voyage dans la Russie méridionale et la Crimée par la Hongrie, la Valachie et la Moldavie* (Paris, 1840), 82.
46 de Perthes, *Voyage*, 398.
47 Quin, *A Steam Voyage*, 18–19.

recalled meetings with "antiheroes," characters who aroused hilarity and made the journey more amusing and thus easier. Among the passengers in the saloon of the steamer on which he sailed from Istanbul to Galaţi were a Moldavian boyar traveling home and a baroness from Pera on her way to Vienna. The discussions between these two—on nobility, history, geography, and religion—were the delight of the whole company. Bolintineanu reports how a "heated and bizarre conversation, conducted in a sort of French, attracted a number of people to listen to it; the laughter was general. From there to Galaţi," he continues, "we had a pastime. Whenever we got bored, we managed to make the baroness start talking with the boyar."[48] Contacts established during the voyage sometimes grew into firm bonds of companionship.[49] Those passengers who had never been in the Ottoman Empire before would gather together and exchange precious information among themselves, or with those who had already visited the Orient. The Austrian Ida Laura Pfeiffer, who set out on her own for Istanbul, recalled that the captain helped her make the acquaintance of a gentleman who "afterwards frequently took me under his protection" in the course of her journey.[50] On his arrival in Galaţi in October 1842 to visit the Moldavian boyar Dimitrie Moruzi, the Swiss botanist Carol Guébhart (1792–?) noted that he was sorry to part with his traveling companions, who were continuing their journey to Istanbul: "On the road, acquaintances are quickly formed, and ten days of life and deprivations together made us almost old friends."[51]

For passengers going upstream, entry to the Habsburg provinces was preceded by a period of quarantine in the lazaretto at Orşova. Andersen traveled from Istanbul to Vienna with a group of Austrian and British officers, and he and the British surgeon and geographer Francis W. Ainsworth (1807–1896) occupied two of the "cells" of the lazaretto. Ten days of sanitary arrest, which I shall discuss in Chapter 5, established a bond between the two men, proving the social benefits of traveling companionship.[52]

48 Bolintineanu, *Călătorii în Moldova*, 264–266.
49 Alasdair Pettinger, "Companion," in Charles Forsdick, Zoë Kinsley, and Kate Walchester (eds.), *Keywords for Travel Writing Studies: A Critical Glossary* (London, 2019), 51–53.
50 Pfeiffer, *Visit*, 19.
51 Carol Guébhart, "Amintiri din Moldova," ed. Marian Stroia, in Daniela Buşă (ed.), *Călători străini despre ţările române în secolul al XIX-lea*, vol. 4, *1841–1846* (Bucureşti, 2007), 179.
52 Andersen, *A Poet's Bazaar*, 166–178.

For travelers covering longer distances, the social activities contributed to the development of a feeling of community. As the German Johann Georg Kohl (1808–1878) explains, "In the steamboats along the Rhine none of this amalgamation takes place, because at each station, the boat takes up as many new passengers as it deposits old ones. Not so on the Danube, where the intermediate stations are as yet of very little importance."[61] As passengers were obliged to live together in steamboat saloons for up to two weeks, the group thus formed becomes an interesting object of historical analysis, visible in various types of interaction, during meals, society games, and sleep.

SLEEP AND ITS ENEMIES

Sleep is increasingly present as an object of research in the social and human sciences. Studies of the sociocultural significance of sleep have tried to place this vital component of human activity in its historical context.[62] As collective dormitories for a cosmopolitan group of travelers, the saloons and the decks of the Austrian steamers constitute a perfect laboratory for a study of the social components of the biological need to rest for at least a few hours per day.

There were sleeping arrangements on all the steamers, but private cabins were rare. First- and second-class passengers slept inside the vessel, while those with third-class tickets rested in the open air or, at best, in tents set up on the deck. The majority of the authors I have examined either shared a cabin or chose to sleep on the deck, and their descriptions of the accommodation occupy memorable pages in their books. With few social spaces available and with the same saloon serving both as day room and as dormitory, sleeping was regulated not only by the biological needs of individual passengers but also by the social norms of the group.

Fleas were numerous, but mosquitoes were a more serious problem in certain periods of the year, as they brought with them the threat of malaria.[63] Travelers came prepared, if possible, with mosquito nets, but the reality was much harder.

---

61 Johann Georg Kohl, *Austria: Vienna, Prague, Hungary, Bohemia, and the Danube; Galicia, Styria, Moravia, Bukovina, and the Military Frontier* (London, 1843), 262.

62 Roger A. Ekirch, *At Day's Close: Night in Times Past* (New York, 2006); Simon Williams, *The Politics of Sleep: Rights, Risks, and Regulations* (New York, 2011).

63 Robert Snow, *Journal of a Steam Voyage down the Danube to Constantinople, and Thence by Way of Malta and Marseilles to London* (London, 1842), 26–27.

The scene recounted by the Frenchman Alexis de Valon (1818–1851) has a particular charm. The passengers had barely settled down to sleep, when

> a buzzing first muted and then louder and louder let us know, as soon as the light was out, that we had another peril to face. There were thousands of mosquitoes in the room. We all got up together yelling as if possessed. The waiters arrived, and presented us with pieces of green gauze that had once been made into mosquito nets; each arranged his veil and we went back to bed. A minute later, the buzzing resumed even louder than before; I felt myself being devoured, and as I kicked the partition in anger, I heard one of my neighbors giving himself a resounding slap. "*Razza del diavolo*!" cried a furious voice: it was the Lazarist [priest]. "*Der Teufel*!" roared the Dutchman. And I leave you to judge whether French oaths were missing from this concerto. At that moment—I shall never forget it—a child started shouting on the deck, and a music-loving sailor tried on his clarinet this air that went on all night without variation: do–mi–re–do—do–mi–re–do. There could be no thought of sleeping.[64]

As for the Scottish writer and diplomat James Henry Skene (1812–1886), he treats the conflict between sociality and biological need in terms of the symbolic confrontation between the day room of whist players and the dormitory of exhausted travelers.[65] As the card games and conversations continued into the night, sleep had to conform to the dictates of social order. Disputes often broke out, but ultimately sleep gradually spread through the common dormitory.

The sleeping conditions also raised issues of social and gender roles, with the result that travelers' accounts are full of amusing stories of what could go wrong in the steamer's collective dormitories. Couples were separated and individuals were supposed to respect the social norms of their gender groups. On the other hand, overcrowding in the women's saloon led some female passengers to seek places in the men's saloons.

In the dormitory, nightwear was one of the sources of culture shock. As the American Noyes notes: "It was amusing to see Turks in twisted turbans, Germans in night-caps, white-coated Austrians, and well-booted Magyars, Servians and

---

64 Alexis de Valon, *Une année dans le Levant*, vol. 2, *La Turquie sous Abdul-Medjid* (Paris, 1846), 188–189.
65 James Henry Skene, *The Frontier Lands of the Christian and the Turk*, vol. 1 (London, 1853), 201–202.

Wallachs, Jews and Gentiles, lying down quietly together, as if the millennium had dawned in the close, dimly lighted saloon of the Albrecht."[66] "When the time of repose arrived," writes Elliott, "each individual undressed as much as he thought fit; the men for the most part but little, the women entirely." He remarks on such breaches of propriety as those of "a girl of seventeen years, with her mother and another female, [who] disrobed themselves in the presence of twenty men, and in the full light of six candles, without attempting to conceal their persons," or of "a mother advanced in years, in the company of her daughter and husband" who "preparatory to a siesta, suddenly stripped herself of very nearly all her clothes but one garment," in a manner suggesting "an utter unconsciousness of indecorum." As for the women's cabin, the intimacy of the occupants sometimes had to be guarded by husbands or other concerned gentlemen. According to Elliott, the British consul to Bucharest, Robert Gilmour Colquhoun, traveling with his mother and sister, "was obliged to turn two men out of the cabin when his ladies wished to repair thither; and his just representation of the inexpediency of their entering it led to a sharp reply in defence of this violation of delicacy."[67]

Crowded cabins made travelers particularly attentive to smells.[68] The situation was especially problematic during the hot days and nights of summer, when the air became unbreathable. When the weather permitted, travelers slept on the deck to escape both mosquitoes and unpleasant odors. Even when it was cold, this was still "a luxury compared with the overheated and crowded cabin."[69]

Passengers often concluded strategic alliances to reserve sleeping places. Oliphant describes in detail the politics of sleep on board the steamer *Boreas*. With the embarkation of more and more passengers at Galați and Brăila

soon the small triangular cabin, miscalled a saloon, is inconveniently crowded. There are only two or three private cabins to be procured, which are pounced upon at an enormous price, and the weather is far too cold to admit of sleeping on deck. The consequence is, that, as night draws on, preparations for turning in are apparent. Those who wish to secure a few

66 Noyes, *Roumania*, 66–67.
67 Elliott, *Travels*, 77–79.
68 Ibid., 72.
69 Brother Peregrine (Octavian Blewitt), "The Danube," part 2, *Fraser's Magazine for Town and Country* 22, no. 132 (1840), 686. On smells, see Vintilă-Ghițulescu, *Patimă și desfătare*, 266–271.

feet of *infested* sofa, wisely take possession about six o'clock. Those who prefer a cigar in the calm moonlight on deck, may esteem themselves fortunate if, upon going below, they can find an unoccupied space on the floor. For about two hours the greatest confusion prevails. Everybody is either fighting for his bed, or making it, or snoring in it, such as it is. Some people do not think it necessary to undress at all; others go to an opposite extreme, and expose themselves to a needless risk of catching cold.

When, having enjoyed the fresh evening air as long as possible, I quit the deck about midnight, I seem to be entering a badly managed hospital rather than the saloon of a steamship. I know my bed is secured to me, because three of us have entered into partnership to watch over one another's interest, and we mount guard alternately.

A Prussian fellow passenger who had not joined this "partnership," however, came back to his bed to find that a "burly Austrian" had taken possession of it. The resulting brawl was stopped by the intervention of two gendarmes, but the incident would be reported to the authorities on arrival in Habsburg territory.[70]

Some travelers dwell on the arrangements for personal hygiene. De Perthes recalled: "There was only one chamber pot, one basin, and one water jug to half a dozen beds. One also had to queue to get a place."[71] Washing was another complicated matter, as Oliphant testifies: "The only basin supplied by the company was required at nine o'clock A.M. by the stewards, so that the crockery might be washed in it immediately after the passengers. It was therefore necessary for some of the party to begin their ablutions before daylight—as we had scarcely done fighting for the basin when we began to fight for places at the breakfast table."[72]

Journeys through Space and Time

One of the characteristics of river travel is the special relation to the historical space and time outside the steamer. Travelers observed at close quarters the natural environment, the economy, and the history of the territory

---

70   Oliphant, *The Russian Shores*, 352–353.
71   de Perthes, *Voyage*, 400.
72   Oliphant, *The Russian Shores*, 353–354.

between Vienna and Istanbul, thus making their journey different from one solely by sea or on the ocean. The voyage was made more interesting by the geography and history of the places they passed, both closely associated with symbolic or memorable spatial and temporal nodes.

Most travel narratives have a relatively similar structure. The text is constructed around two complementary realities: an "outside world," that is, the spatial-temporal dimension along which the steamer and its passengers move, and an "inside world," that is, the social environment associated with the cosmopolitan community the traveler joins upon embarking on the steamer. In this chapter, I insist on this second aspect, which enables a better understanding of the early stages in the industrialization of travel and the new social bonds favored by the steamer. More than the "outside world" through which the vessel was passing, this inner social micro-universe is central to each traveler's narration and individualizes their personalities and ideas.

The "outside" spatial-temporal dimension is a supplementary projection of this "inside world." Traveling through space-time and getting to know the various places of interest along the Danube was a social act, mediated either by written sources (travel guidebooks or maps) or by interactions with other passengers who possessed local knowledge of the geographical spaces and historical times that were relevant at each place and moment in the course of the voyage. Elliott refers to such a case, when the "native gentlemen" on board the steamer "pointed out every object of note in our route, furnishing the name and history of each successive locality."[73] Further downstream, "each turn in the river had presented a fresh subject for some amusing anecdote, some historical allusion, or some political opinion" to be shared among the travelers.[74] As in the case of the Hungarian poet Ferenc Kazinczy (1759–1831), the journey was not a mere linear movement through geographical space, but an intense synthesis of multiple contexts linked to the traveler's own observations, his or her experience of life, and the experiences of the other travelers.[75]

---

73 Elliott, *Travels*, 48.
74 Ibid., 51.
75 Wendy Bracewell, "Travels through the Slav World," in Bracewell and Alex Drace-Francis (eds.), *Under Eastern Eyes: A Comparative Introduction to East European Travel Writing in Europe* (Budapest, 2008), 147–194, n. 18, and Irina V. Popova-Nowak, "The Odyssey of National Discovery: Hungarians in Hungary and Abroad, 1750–1850," in Bracewell and Alex Drace-Francis (eds.), *Under Eastern Eyes: A Comparative Introduction to East European Travel Writing in Europe* (Budapest, 2008), 195–222. For references to the Hungarian experience, see Tinku-Szathmáry, "Gőzhajóval," 11–38.

The voyage between Vienna and Istanbul was an instructive one for travelers with an interest in European history.[76] Between Vienna and Budapest, the banks of the Danube were "monotonous and destitute of picturesque beauty, but historically interesting." Travelers could admire the site of Rudoph of Habsburg's victory over Ottokar of Bohemia, a foundational event in the history of the Austrian Empire, or could, in their minds, meet Napoleon at Aspern, Essling, and Wagram. A stone monument outside the village of Schwechat marked the spot where Emperor Leopold met John Sobieski after the Polish king and the duke of Lorraine had ended the second siege of Vienna in 1683. Further downstream, Petronell stood on the site of the Roman settlement of Carnuntum, sacked by Attila, and a Romanesque chapel there was thought to have been built by Charlemagne or the Knights Templar. At Heidentor, travelers could admire the ruins of a triumphal arch believed to have been erected by Augustus to mark the conquest of Pannonia by the future emperor Tiberius. Pressburg (Bratislava) was a town with symbolic significance for the Hungarian constitutional system, as the seat of the Diet and the place of coronation of kings of Hungary, and it brought to mind Maria Theresa's famous meeting with the Hungarian nobles. After Bratislava, the Danube spread out into several "arms," the principal of which took travelers downstream to Martinsberg, the site of the splendid Benedictine Abbey of Saint Martin (Sz. Marton), the oldest monastery in Hungary, founded at the end of the tenth century by Géza, the father of Saint Stephen. An indecisive battle took place at Ács in July 1849 between Hungarian patriots led by Artúr Görgei and Austrian imperial forces under Julius Jacob von Haynau. The fortress of Komárom/Komárno, situated at the southern end of the island of Schütt, was one of the strongest in Europe. Its foundation was linked to the name of King Matthias Corvinus, but it had been destroyed in the eighteenth century by an earthquake, and thus prompted discussion of the great energies released by such dramatic upheavals, whether geological or political in nature. Esztergom was the episcopal see of the primate of all Hungary and the location of a magnificent cathedral, while the emerging capital of Hungary, Budapest, had its own fascinating history too.

---

76 The presentation that follows is based on *A Handbook for Travellers in Southern Germany*, 8th ed. (London, 1858), 496–539. For a comparative perspective on later tourist literature, see Ploner, "Tourist Literature."

In the next section of the journey, the river crossed the "European pampa," a vast plain where travelers could admire the rich agricultural land and see sites recalling confrontations between Christianity and Islam. One such was Mohács, famous for the battle in 1526 at which Suleiman the Magnificent put an end to the independence of the medieval Hungarian kingdom. The fortress of Petrovaradin (part of today's Novi Sad) was said to have been named after Peter the Hermit, who assembled the forces of the First Crusade there, while Karlowitz (Sremski Karlovci) reminded passengers of the peace treaty signed there in 1699, by which the Habsburg monarchy obtained the provinces of Hungary, Slavonia, and Transylvania.

Semlin (Zemun) and Belgrade marked the beginning of the gradual passage toward another cultural world, one in which church spires and minarets faced each other from opposite banks of the Danube. The citadel of Belgrade was the legacy in stone of John Hunyadi, the defender of Christendom in the fifteenth century. The city also evoked the history of Serbia's struggle for autonomy, while Passarowitz (Požarevac) was the site of the signing of the peace treaty of 1718, after Prince Eugene of Savoy's brilliant victories. The feudal castle of Golubac was, according to popular belief, the place where Saint George slew the dragon. Further downstream, at Turnu Severin, passengers could admire the remains of a marvel of ancient engineering, the bridge "which time, violence, and the floods and ice-shocks of 1600 winters, have not been able to destroy [...] built, A.D. 103, by the architect Apollodorus of Damascus, who also erected Trajan's column at Rome after the defeat of the Dacian King Decebalus." On the right bank, the citadel of Vidin guarded the border of the Ottoman Empire, while Nicopol (Nikopol) recalled the epic battle in 1396 between Sigismund of Hungary and Sultan Bayezid.

Closer to the mouths of the Danube, a new empire—Russia—began to make its presence felt in the geographical and historical landscape. Reni was the first town of the Russian Empire on the Lower Danube, but it was Izmail, which featured in Cantos VII and VIII of Byron's *Don Juan*, that most attracted the interest of travelers. In the Black Sea, the Greek heritage was visible on the island of Leuke (Snake Island), where the ruins of a temple dedicated to Achilles could still be seen. The ancient Greek colonies of Tomis (Constanța) and Odessus (Varna) bore the marks of the recent conflicts between the Russian and Ottoman Empires, but the most fascinating place of all was Istanbul itself, with its multimillenary heritage.

In their writings, travelers mixed historical evocations with details about the economy and ecology of the region, which was rich in vineyards, fish, and birds. More than any other means of communication, the steamer allowed passengers to move through space-time with a 360° view of the surroundings and the explanations of improvised guides. Most of their narratives involve a continuous oscillation between cabin life and the "outside world." "In default of interest in the passing country," remarks the German geographer and historian Kohl, "I turned my attention to the little community around me."[77]

CONCLUSIONS

As a global phenomenon, travel meant more than modern infrastructure, fast vehicles, and massive power of propulsion. It came also with an inherent social dimension that evolved concomitantly with the modernization of public transport. The transport revolution of the nineteenth century not only brought higher speed, greater comfort, and prices to suit all pockets, it also gave rise to new forms of social contact for the passengers who embarked on a steamboat or took their seat in a railway carriage.

In line with new directions of study in the history of transport,[78] this chapter has included the steamboat in the emerging model of mobile sociality. The Austrian steamer served as an interimperial contact zone, slipping between a number of different worlds, and serving as a space of connection between two of the world's most attractive cities. According to the various contemporary accounts, the cabins and decks of the steamboats were veritable micro-worlds, in which global actors met as a result of their need to travel. For most of the writers, the cabin was the center of this universe, where passengers with greater power from a financial point of view met, interacted, and often formed a mobile community for which sociality was an integral part of the experience of travel. Living in this physical space meant not only the exchange of precious information about the various stages of the journey but also the sharing of diverse social norms and practices.

---

77 Kohl, *Austria*, 251.
78 Massimo Moraglio, "Seeking a (New) Ontology for Transport History," *Journal of Transport History* 38, no. 1 (2017), 3–10.

The Austrian steamboat became an effervescent space of global connections and of sociality, a perfect example of the manner in which mobility and sociality combine and transcend the already flexible boundary between private and public space. Detained on board for a period of several days at the very least, travelers explored new "mobile temporalities,"[79] fixed in the cosmopolitan community's ad-hoc practices, which ranged from the hour of dinner to that of sleep and which pushed passengers to adapt to the conviviality of their vehicle. Seen through the prism of such personal and subjective accounts, the vessel is a heterotopia, a fascinating stage for global history, a spatial, cultural, and social connector and mediator of the world.

---

79   Nicola Green, "On the Move: Technology, Mobility, and the Mediation of Social Time and Space," *Information Society* 18, no. 4 (2002), 281–292.

CHAPTER 3

Between the Orient and Russia

ROMANIAN IN-BETWEENNESS

The history of the "Carpathian–Danubian–Pontic" space is often explained in terms of its characteristic of being "at the crossroads of civilizations."[1] Various authors have examined this geographical, political, economic, or cultural positioning between civilizations or empires. The most obvious historiographical placing is that suggested in the title of Neagu Djuvara's book: "Between Orient and Occident."[2] These two vectors are valid if we consider the political dependence of the Principalities of Wallachia and Moldavia on the Sublime Porte, or their traditional culture and Byzantine-rite religion as representing the "Eastern" dimension of Romanian history, but also the aspiration to Western modernity and a national identity centered on Latinity. With particular reference to economic matters, Bogdan Murgescu places the medieval and premodern Romanian space "between the Ottoman Empire and Christian Europe."[3] Situated "in between" different geopolitical structures and imperial forces, the Principalities were marginal spaces, lying on the periphery of the Ottoman Empire or on the borders of Europe.

Various travelers of the nineteenth century seem to have accepted such a placing. Princess Aurélie Soubiran Ghica (1820–1904) considered Wallachia "a stopping place between East and West." It was a territory full of contrasts, but which had managed to preserve its specific character in an age in which modern

---

1   Lucian Boia, *Romania: Borderland of Europe*, trans. James Christian Brown (London, 2001), 11.
2   Djuvara, *Le pays roumain*.
3   Bogdan Murgescu, *Țările Române între Imperiul Otoman și Europa creștină* (Iași, 2012).

civilization—globalization, we would say today—was leveling the differences between the countries of the world.[4] James Henry Skene, who had been resident for some time in southeastern Europe and had a good knowledge of the region, located the Principalities, together with other parts of the Balkans, in the broader region of "the Frontier Lands of the Christian and the Turk."[5]

Historians have examined such geographical and cultural positionings, the result of the growing international interest in the Principalities of Moldavia and Wallachia starting in the second half of the eighteenth century. However, the accounts written from the 1830s to the 1850s by passengers on the Austrian steamers make reference to another determining special axis for the history of the Principalities, seen primarily as an intermediary space between the Orient and Russia, in which the West (or "Europe") was barely beginning to make its presence felt. The introduction of a steamboat service between Vienna and Istanbul contributed decisively to the mapping, both literally and in a symbolic sense, of Central and southeastern Europe. While in the Habsburg territories the mighty river was an important element of economic and cultural cohesion, a spinal column or main artery of the "Danubian monarchy," the lower reaches of the Danube had become an interimperial border delimiting spaces, political entities, and fluid identities. As will be discussed in more detail in Chapters 5 and 6, the Danube Gorge (and especially the Iron Gates) marked the geographical and symbolic passage between West and East, at the end of a space of transition that had already begun at Belgrade. Downstream from the natural barrier of the Carpathian Mountains, the Danube separated the Ottoman Empire from the Principalities of Wallachia and Moldavia and then from Russia along a border established by the Treaty of Adrianople (1829). Along the entire lower course of the Danube, from Turnu Severin to Sulina, travelers were witnesses to the struggle, but also to the cohabitation of the two rival powers: imperial Russia and the Ottoman Empire. As seen by these wanderers, the region was marked by a mixture of the decay and exoticism specific to the Turks and perfidiousness and aggression characteristic of the Russians, features seen through the prism of all the Orientalism and Russophobia of the age.

4   Aurélie Ghika, *La Valachie moderne* (Paris, 1850), 25.
5   Skene, *The Frontier Lands*, vols. 1–2 (London, 1853), published anonymously as "by a British resident of twenty years in the East." A few years later, the American James O. Noyes chose a similar title for his *Roumania: The Border Land of the Christian and the Turk* (New York, 1857).

This chapter is about perceptions, clichés, and the imaginary. It aims to examine the way in which the Principalities of Wallachia and Moldavia took shape in the descriptions of writers who had "got to know" the lands north of the Danube mainly at steamboat speed. Their travel accounts include ample references to the two Principalities' political, economic, and social land-scape, and in the following pages, I shall synthesize some of the recurrent themes in their writings. These are based partly on empirical observations—starting from Moldavians and Wallachians on board, local people at ports along the way, short visits to Giurgiu when the steamer stopped there or to Galați during the transfer from the river steamer to the sea-going vessel—but also, above all, on clichés recycled from guidebooks and travel literature the writers had read.

On the basis of these sources, the image of the Principalities took shape in a similar manner to the way other discursive spaces in the geographical vicin-ity were invented. The classic works of Edward W. Said,[6] Maria Todorova,[7] and Larry Wolff[8] are very relevant here, especially inasmuch as the territory north of the Danube was constructed in the Western imaginary out of a combination of clichés examined in each of their studies, together with ste-reotypes deriving from the Russophobia of the majority of the Westerners who traveled along the Danube in the period leading up to the Crimean War.[9] It is interesting to trace the forms and doses in which these images are superimposed and how symbolic geography is sketched at steamboat speed, both through the hybridization of certain clichés specific to the empires that controlled the two Principalities and through the appearance in the equa-tion of the aspirations of the Romanian nation in the making.

SPIRES AND MINARETS, TOWERS AND RUINS

The quarantine cordon established along the Danube—with lazarettos in the ports and hundreds of border guard posts spread out along a distance of more than a thousand kilometers—was the visible, institutionalized form of the Danube frontier. Taking the form of rudimentary structures raised on

---

6   Edward W. Said, *Orientalism* (London, 1978).
7   Maria Todorova, *Imagining the Balkans* (Oxford, 2009).
8   Wolff, *Inventing Eastern Europe*.
9   J. H. Gleason, *The Genesis of Russophobia in Great Britain 1815–1841* (Cambridge, MA, 1950).

tall posts in order to be safe from variations in the level of the river, the guard posts not only made up part of the sanitary "dyke" but also contributed to the picturesque character of the region and to the distinctive identity of the lands to the north of the river.

The preponderantly political function of the quarantine is, as will be shown in more detail in Chapter 5, a leitmotif of the travel literature. In the view of many passengers, the true role of the cordon sanitaire was to obstruct connections between the Ottoman Empire and the Principalities, in preparation for the moment when Russia would be able to annex them without too much international protest. This was only an intermediary step toward conquering the Balkans and obtaining the much-desired prize: control over the Straits of the Bosphorus and the Dardanelles. Having once served as the Roman Empire's shield against the barbarians of antiquity, the Danube now played the same role against the barbarians of modern times—the Russians— in the view of the Frenchman Jacques Boucher de Perthes, who traveled on the river in 1853, when the Russian—Ottoman conflict was smoldering.[10] As a sanitary and a military frontier at the same time, the Danube had acquired a clear identity in the political and symbolic geography of the region: it was the barrier by means of which Russia defended itself against the plague, but also that along which the Sublime Porte protected itself from the territorial insatiability of Russia.

In a space with a fluid political status, the Danube flowed between two "savage banks," but with different types of barbarism, which could be clearly distinguished from the deck of a steamer. Seen from the thalweg of an international waterway, the two banks looked different not only politically, historically, and culturally but also geologically. The Bulgarian bank is higher, and the string of Turkish citadels downstream from Vidin made it seem relatively well fortified (Figures 6 and 7). The Wallachian shore is lower, offering a view of floodplain and thinly populated and little cultivated lowland.

The pairs of towns on the opposite banks illustrated the differences even more clearly, highlighting the military, cultural, and religious particularities of the two territories. On the way downstream, a traveler could see on the right bank the citadels of Vidin, Ruse, and Silistra, and on the left the market towns of Calafat, Giurgiu, and Călărași. The Bulgarian bank seemed more

---

10 de Perthes, *Voyage,* 373.

Figure 6 William Henry Bartlett, *The Balkans [from near Vidin]* (c.1840).

Figure 7 Bartlett, *The Plains of Lower Wallachia [from the Castle of Sistova]* (c.1840).

exotic, with its citadels and minarets, with tales of heroism and memorable legends. The American physician Valentine Mott (1785–1865), approaching the end of a long tour of Europe and the Near East in 1841, noted these differences, as far as he could perceive them from the deck of a steamer sailing upstream on the Turkish side:

> The numerous minarets and mosques on the Turkish banks strikingly contrast with the more humble and unobtrusive Christian temples on the opposite side of the river.
> Contrary to our expectations, the Turkish territory seemed evidently to present, in its advanced state of agriculture and general appearance of comfort, a much higher degree of civilization and social improvement than had been attained by their Christian neighbours.[11]

Sailing downstream on the Wallachian side in 1850–1851, the Scotsman Skene made similar observations. On the one side was "the fortified town of Widin"; on the other was "the straggling village of Calafat": "The minarets and cypress-trees of the former offered a striking contrast with the bare and wretched appearance of the latter." Further downstream, Nikopol, the site of a famous confrontation between cross and crescent, with its "clusters of white houses and shining minarets, perched on a line of limestone cliffs," contrasted with Turnu Măgurele on the Wallachian side, "a low and miserable-looking place," and Svishtov, where the peace treaty of 1791 had been signed, ending another confrontation between empires, contrasted with the Wallachian market town of Zimnicea. On the Wallachian side the view was now of spreading villages "lying on the open plain, so flat and bare, that the high and wooded country on the southern side, though little in itself, seemed quite picturesque in comparison" (Figure 8). There followed Ruse (Figure 9) on the right bank, with its "strong military works" and a garrison of thousands of soldiers, and Giurgiu on the left, whose fortifications had been dismantled by the Russians.[12]

For Laurence Oliphant too, as he sailed upstream on the Wallachian side in 1852: "As usual, one side monopolises all the beauty; and the picturesque

---

11   Valentine Mott, *Travels in Europe and the East* (New York, 1842), 445–446.
12   Skene, *The Frontier Lands*, vol. 1, 198–206.

Figure 8 Bartlett, *Sistova, from the Turkish Cemetery* (*c.*1840).

Figure 9 Bartlett, *Rutzscuk* (*c.*1840).

Turkish towns, with their mosques perched upon the steep hill-sides, or peeping out from amid woods and vineyards, cause those passengers who are susceptible of them, passing emotions."[13]

The image was as clear as it could be: citadels and minarets on one shore, market towns and church spires on the other. The Muslim bank was strengthened with fortifications (albeit not in the best state of repair), while the Christian side could show only the ruins of fortifications demolished on the orders of Russia, which in the eyes of many passengers was the occult power that stalked the Danube Plain, ready to extend its malefic domination across the river too.

Beware of Russians, Even Bearing Gifts

As the border that separated the Principalities from the suzerain power, the Danube was a river of "salvation" for Wallachia and Moldavia. Freedom of navigation was a result not only of the technological revolution but also of major political decisions. The great river had been "freed" by the provisions of the Treaty of Adrianople, under which restrictions on the Principalities' foreign trade were removed. After 1829, the Danube had thus become one of the principal routes toward the liberation, not only economic but also political, of the territories on its banks.

This special relationship with the river turned Moldavia, Wallachia, and Serbia into the "Danubian Principalities," an expression which later came to refer only to the first two. The term had limited circulation prior to 1830. It was in the period of the 1848 revolutions and above all the Crimean War that it came to be used more intensely, when the Danube question and that of the union of Moldavia and Wallachia became more pressing and internationalized. As I explained elsewhere,[14] an important component in the creation of modern Romania concerned its mission as guardian of freedom of navigation on the Danube, an important principle included in European law under the Treaty of Paris (1856).

For small states like the Principalities of Moldavia, Wallachia, and Serbia, their economic future was closely bound to the Danube, a vital

13 Oliphant, *The Russian Shores*, 351.
14 Ardeleanu, *The European Commission of the Danube (1856–1948): An Experiment in International Administration* (Leiden, 2020).

transport infrastructure connecting them to global commercial markets. However, Russia, the same empire that had made a decisive contribution to their economic liberation, held control of this strategic highway, which started (or ended) in the swamps of the Danube Delta, a territory annexed by the Russians in 1829 without any great protest on the part of the Western powers.

Russia is the ubiquitous power in the literature of travel on the Lower Danube in the two decades preceding the Crimean War. Exercising control openly or discreetly, Tsar Nicholas I was presented as all-powerful in this region too. For passengers coming from Istanbul, the presence of Russia made itself felt already at Sulina, where the travelers were witnesses to what they considered characteristic manifestations of Russian power—negligence, corruption, and abuse—which sometimes also affected the circulation of the Austrian steamers. For Oliphant, the Carpathian Mountains in the region of the Iron Gates were "the present limit of Russian aggression."[15] For travelers going downstream, Russia entered the discussion whenever the situation of Moldavia and Wallachia, seen as the first victims of any anti-Ottoman plan on the part of the Russians, was presented. Russia became more and more visible as travelers approached the Black Sea, a maritime space which it dominated strategically in confrontation with the Sublime Porte.

The majority of Western passengers shared the Russophobic vision in which Russia was considered a despotic empire, eager to subjugate both the Danube and the territories on its shores. However, each author brings a different accent, with the British leading the way in the circulation of Russophobic clichés. The new international status of the Principalities indicated, in a far-from-subtle manner, who their real master was. For the Englishman Charles B. Elliott, "formerly of the Bengal civil service," who traveled on the Danube in 1835, the removal of the Phanariot rulers and the enthronement of the new "hospodars" ("a word corrupted from the Russian *gospodin*, lord") had simply been a transfer of authority to Russia. The Porte's suzerainty was purely formal, as Russia held absolute control and nothing could be done without its approval. Elliott cites as an example the dispute over the additional article to the Organic Regulation of Wallachia.[16] The consul communicated to prince

---

15 Oliphant, *The Russian Shores*, 359.
16 The Organic Regulations were proto-constitutional laws enforced in the Principalities in 1831–1832, during a period of Russian military occupation (1828–1834).

Alexandru Dimitrie Ghica (1796–1862, reigned 1834–1842) the displeasure of Tsar Nicholas I, and Ghica responded with a letter "expressive of his regret that he should unintentionally have given umbrage to the emperor." This was not sufficient, however, and the consul insisted that Ghica should apologize in person. "Accordingly," Elliott concludes, "the prince of Wallachia was actually seen a suitor for pardon at the door of a Russian employé!"[17] Frances Anne Vane, marchioness of Londonderry, traveling in the region in 1840 with her husband, an influential British diplomat, similarly notes that the Principalities "are nominally protected by Russia, a significant term, importing that they will soon belong to her."[18]

For Oliphant, the situation of Bessarabia was a model of bad practice, an image of how the future of the Principalities would look if they were annexed by Russia. Referring to the difficult situation of this province under the tsar's rule, he makes gloomy predictions:

> Should the Emperor grant them a constitution, they can compare it to that which Alexander granted to the Boyars of Bessarabia, and need be under no uncertainty as to the extent of its duration. Should he accord them special privileges, they will at once be able to estimate them at their true value, to anticipate their fatal effects, and to calculate exactly how long it will be before protection in trade shall reduce them to a state of Bessarabian depression.[19]

The most malefic form in which Russian interference manifested itself was to be seen in the Danube Delta. In controlling the access to and from the Black Sea of the provinces on the shores of the Danube, Russia in fact controlled their prosperity. The imperial authorities denied any involvement in the obstruction of navigation on the river, but their gunboats at Sulina enabled them to intervene whenever necessary. The British officer James John Best (?–?), traveling on the Danube in 1839, described this potential power of Russia to turn off the tap of Danube trade, announcing hard times for the prosperity and indeed survival of the territories upstream.[20] Thus, far

17 Elliott, *Travels*, 156–158.
18 Vane, *Narrative*, 81.
19 Oliphant, *The Russian Shores*, 340–342.
20 J. J. Best, *Excursions in Albania* (London, 1842), 295.

from being the "liberator" of the Principalities, Russia had, under the terms of the Treaty of Adrianople, found new ways in which to control Wallachia and Moldavia while waiting for the occasion to annex them.

THE SHIFTING SANDS OF SULINA

When the Irishman Patrick O'Brien (1823–1895) arrived at Sulina in September 1853 on board the steamer *Ferdinand I*, he found "something fearfully desolate" in the landscape around the river mouth. Wrecks could be seen protruding from the murky tide water. "Stranded on the shore," writes O'Brien, "was the large hull of a Dutch-built vessel, rotting in the sun, and close to us were some men in boats, trying to fish up the cargo of a vessel which had gone down the day before." The steamer was unable to enter the river, as the depth of the navigable channel over the bar—the sandbank that forms naturally where a river enters the sea—was insufficient to allow safe passage. The passengers were taken over the bar in a barge and transferred at Sulina onto the river steamboat that was to take them upstream toward Galaţi and Brăila.[21]

The predictions of those writers who for twenty years had been warning that Russia would block navigation seemed to be coming true. The mouths of the Danube were being closed by gates of sand that limited access to the increasingly prosperous commercial outlets of the Principalities. The problem of Sulina as a diplomatic dispute between Russia and the great powers with an interest in Danube navigation—about which I have written at greater length elsewhere—was at its apogee.[22] The Austrian Lloyd company was directly affected by the depth of the Sulina mouth, and the solution mentioned by O'Brien—transferring passengers over the bar by barge—was designed to prevent further accidents. For commercial vessels, however, such a solution was both dangerous and costly.

The Venetian prelate Francesco Nardi (1808–1877), professor of theology at the University of Padua, had already captured the suspense of the passage over the fateful sand threshold, the unacknowledged "work" of Russia. In the autumn of 1852, Nardi was traveling from Trieste to Istanbul, via Vienna.

---

21 Patrick O'Brien, *Journal of a Residence in the Danubian Principalities in the Autumn and Winter of 1853* (London, 1854), 7–8.
22 See Ardeleanu, *International Trade*.

When the *Ferdinand I* arrived at Sulina, a Russian boat approached the steamer to check documents and to announce the depth at the bar: 9½ feet. The steamer needed a minimum of 8¾ feet to pass safely, but

> [the captain placed at the prow of the vessel] two sailors who were to sound the water with the help of rods, where the current was very strong. The captain exchanged a few words with the pilot, and then gave orders to the engineers: full steam ahead! Silence reigned all around; only the sound of the paddles could be heard. We were soon near the dangerous sandbank, and the men who were sounding called out the depth of the water: 14, 12, 11, 10, 9; immediately there was a yell. We had touched the bottom, but the vessel was immediately afloat again [...] while 8½ was called. Either the gentlemen on the boat were misled or they wanted to mislead us. We were in a dangerous situation; in the distance the cemetery of ships could be seen, spreading on either side of the dismal river. Masts rising out of the water, overturned keels, timbers half buried, that was the port of Sulina. But the man taking soundings began to call out again: 12, and then the depth rapidly increased to 20, 25, 30, and at 32, the sounding was abandoned, as it was of no further use. The captain came up to us happy and said, "We are out of any danger."[23]

Sulina was the key to the Danube. The town had grown together with the sandbank, which had, the majority of ship-owners, merchants, and consuls in the ports of Brăila and Galați believed, become larger due to the calculated indifference of Russia. Sulina had acquired its prosperity on the back of the vessels that had to wait for sufficient depth of water to allow entry to the river or passage out into the Black Sea. Many ships transferred their loads onto smaller vessels in order to be able to navigate the sandbanks, the most feared of which was the Sulina bar. This operation—known as lighterage in nautical terminology—gave rise to new opportunities for theft, fraud, and blackmail. Foreigners alleged that these forms of piracy and banditry were part of the deliberate policy by which Russia sought to discourage the

---

23   Francesco Nardi, *Ricordi di un viaggio in Oriente* (Roma, 1866), 40–41.

Figure 10 Bartlett, *Sulina, Mouth of the Danube* (*c.*1840).

commerce of Brăila and Galați to the advantage of its own ports. The officers in Sulina were the big winners, as they hired out their own vessels for
the operation of lighterage and charged considerable fees for this. The officers and imperial administrators in Sulina, it was further alleged, turned a
blind eye or even took part in the acts of banditry that had made the crossing of the bar a difficult and costly operation (Figure 10).[24]

In short, the accusation was that, taking advantage of natural variations
in the sand bar, the Russians were obtaining considerable material profit
through the raising of new artificial barriers in the way of free navigation.
However, as the British officer Adolphus Slade (1804–1877), later famous as
an admiral in the Ottoman navy, appreciated, the importance of Sulina was
not merely economic but also political. Control of the mouths of the Danube
gave Russia "the power of exercising a direct control over Wallachia and

---

24 TNA, FO 195/136, fol. 538 ("Report on the Navigation of the Danube," drafted by Charles Cunningham, Galați, February 6, 1840).

Moldavia." At some point, the sandbank would provide the perfect argument "to make the Moldavians and Wallachians incline to a junction" with Russia—in other words, to accept the annexation of the two Principalities.[25]

That this point had come in 1853, when Russia occupied Moldavia and Wallachia and the Sulina bar had become almost impassable, was also Oliphant's belief. He synthesizes, in as simple terms as possible, the fears of merchants with regard to navigation at the mouths of the Danube: Russia had aimed for two decades to close the Sulina branch and had neglected all its obligations, assumed under international treaties, to carry out the hydraulic works necessary to keep it navigable. "If the Soulina should silt up," he concludes, "it is probable that the Kilia branch would again be opened, and the fortress of Ismael [Izmail] would command the trade of the Danube."[26]

Before being suffocated by such diabolical schemes, Russian Sulina had seen a period of explosive development, becoming a cosmopolitan and prosperous town. Travelers left various descriptions of the (literally) captivating settlement, about which I have written elsewhere.[27] Steamboat passengers obtained their information from various sources, ranging from details picked up from crew members[28] and local merchants on board to articles in the press of the time (which gave extensive coverage to the Sulina question) and accounts in travel literature and guidebooks. The 1837 Murray edition, for example, not only mentions the quarantine station at Sulina but also the fact that alluvial deposits were increasing the risk of a complete blockage of navigation. With Russia "mistress of the entrance to the Danube," the European powers had to keep watch lest the river be closed by such obstacles, at the same time natural and artificial.[29] The 1853 edition notes the political importance of Sulina and mentions the agreement signed by Russia and Austria in 1840, by which the Russian authorities undertook to carry out hydraulic works to keep the channel navigable in exchange for the payment of a tax by vessels under the Austrian flag.[30]

For most Western passengers, Sulina, according to the Russophobic stereotypes of the period, was illustrative of all that defined the tsarist

25 Slade, *Travels*, 194, 205.
26 Oliphant, *The Russian Shores*, 347.
27 Ardeleanu, *International Trade*, 216–225.
28 Vane, *Narrative*, 92.
29 *A Handbook* (1837), 392–393.
30 *A Handbook for Travellers in Southern Germany*, 6th ed. (London, 1853), 548–550.

regime: militarism, autocracy, egoism. At the same time, it was the example by which Tsar Nicholas I had proved his diplomatic mastery: in annexing the insalubrious and uninhabited marshes of the Danube Delta, Russia had in fact obtained the key that could lock or unlock trade—that is, prosperity—along one of the great rivers of Europe.

PLANS FOR A CANAL

As a political solution to the problems of navigation through the Danube Delta seemed hard to find, a number of authors, including many steamboat passengers, proposed the use of modern technology to establish a route that would be independent of tsarist schemes. If it was not possible to remove Russia from the Danube, then perhaps the Danube, or at least its flourishing trade, could be removed from the Russians.

The idea of constructing a canal in the narrowest part of Dobrogea—between Cernavodă or Rasova and Constanța—was frequently mentioned from the 1830s onwards. Integrated in the route between Vienna and Istanbul, a canal of some 60 km would avoid a circumnavigation of 400 km, part of it through Russian territory controlled by gunboats and by the fortifications at Izmail, thus greatly shortening the journey between the two capitals.[31] The DDSG management explored the idea, and diplomats from Vienna discussed it with representatives of the Porte. A number of engineers visited Dobrogea and carried out what today would be called feasibility studies. From a technical point of view, the specialists considered, the canal could be constructed. However, they drew attention to certain engineering problems—principally regarding a series of elevations along the route—which would increase the costs involved, probably beyond the capabilities of a company like the DDSG.[32]

After negotiations with the Ottoman Empire, the Austrians obtained permission to transfer passengers and goods overland, along a road crossing Dobrogea between Cernavodă and Constanța. A number of logistical provisions along the way, which I mentioned in Chapter 1, enabled the route to become operational from 1839. Over the next four years or so, numerous

---

31 David Urquhart, *Turkey and Its Resources* (London, 1833), 166–167.
32 See, with relevant modern literature, Ardeleanu, *International Trade*, 185–196.

travelers[33] crossed the province and mentioned in their accounts of their journeys the advantages that would result from the proposed canal.

One of them was the English physician and geographer Francis W. Ainsworth (1807–1896). After taking part in many research expeditions in the Orient, he was making his way to Vienna in the spring of 1841, as part of the group that included Hans Christian Andersen. In an article published in a British magazine, Ainsworth put forward his opinion about the possibility of constructing a canal between the Danube and the Black Sea, as a solution to the problem of river navigation through Russian territory. The presence of several lakes and a water course that seemed to communicate between them along a large part of the Dobrogea route, together with the existence of a raised terrace, considered to be part of a canal made between the Danube and the sea by Emperor Trajan, made him a supporter of the project. In any other country, Ainsworth believed, the project would have been carried out long ago. He reinforced his opinion with arguments based on geology, his own area of competence. Including also a sketch map of Dobrogea, Ainsworth concluded that all that was needed to solve the problem of Danube navigation was initiative and moderate expenditure, especially as the geological structure seemed to indicate that a branch of the Danube had at one time crossed the region.[34] It may have been as a result of a discussion with Ainsworth that Andersen also shows an interest in the matter: "During the whole of our day's journey, the lake of Kurasu [Carasu], which is said to be the remains of the canal by which Trajan united the Danube and the Black Sea, lay on our left. It would be an easy matter to repair the damage, yet it would be less expensive to lay down a railway on this level extent of land."[35]

As the passage of the Sulina bar became increasingly difficult in the 1840s, the canal project acquired more and more supporters. Various writers mention the Austrian–Ottoman negotiations on this theme, which ended with the sultan's refusal to permit the construction of works of infrastructure. It was, they believed, the intrigues of the tsar, making use of intimidation and

---

33 TNA, FO 78/363, fol. 28–30 (R. G. Colquhoun to Palmerston, Bucharest, April 12, 1839), fol. 110–115 (Colquhoun to Palmerston, Bucharest, October 17, 1839) and fol. 129–131 (Colquhoun to Fox Strangways, Bucharest, November 9, 1839).

34 Francis W. Ainsworth, "The Communication between the Danube and the Black Sea," *Mirror of Literature, Amusement and Instruction* 10 (September 1842), 152–154.

35 Andersen, *A Poet's Bazaar*, 108.

corruption, that had led to this result,[36] as Russia could not accept losing control of navigation through the mouths of the Danube.[37] Thus, for supporters of the project, the canal meant liberation both from the constrains of the natural environment (the marshes of the Danube Delta) and from Russian scheming.

The technical and economic feasibility of the project resulted in equal measure from both geographical and historical factors. From the geographical point of view, proponents evoked the existence, between Cernavodă and Constanța, of natural lakes that would facilitate the construction works and, above all, the fact that a dried-up branch of the Danube lay in the vicinity of Constanța. From the historical point of view, belief in the engineering competence of the Romans and the existence of Trajan's Wall indicated that all that was needed was to remake what had already existed in the civilized times of antiquity.

Travel guides also gave considerable space to the subject. The 1850 edition of Murray mentioned, with reference to the opinions of engineers, the difficulties involved in constructing a canal.[38] However, the majority of travelers who referred to the canal were convinced of its feasibility. The Italian Nardi reminded his readers that engineers could make use of

> a series of lakes that the Turks named *kara-su* and the Bulgarians *cernavoda*. They could then use the famous Trajan's Wall, which crossed land that was almost a plain, as the few heights at Babadag were lost among the wide valleys. Thus, it would be possible to avoid the mouths of the river, and the journey from Vienna to Constantinople could be shortened by two days. Unfortunately, however, the project remains only a forlorn hope, as no government, or even a private individual, would venture upon such costly works close to that so troubled frontier.[39]

Oliphant considered that the canal would be "a work of great comparative facility,"[40] and the French historian Théophile-Sébastien Lavallée

36 TNA, FO 78/336, fol. 136 (Colquhoun to Palmerston, Bucharest, September 20, 1838).
37 Urquhart, *The Mystery of the Danube* (London, 1851), 109.
38 *A Handbook for Travellers in Southern Germany*, 5th ed. (London, 1850), 520–521.
39 Nardi, *Ricordi*, 29.
40 Oliphant, *The Russian Shores*, 349.

(1804–1866) recalled, also in 1853, that apparently the Danube had once flowed into the sea at Constanța. The proposed construction of a canal, "which would have had such considerable results, especially since the mouths of the river belong to the Russians," had not been put into effect, but it might be replaced by a railway.[41]

The project was revived in the context of the Crimean War, when, in 1856, the British engineer Thomas Wilson obtained the agreement of the Sublime Porte for the construction of a canal between the Danube and the Black Sea.[42] Rival investors stole the march on him, however, and convinced the government that a railway would be more appropriate to the needs of the Ottoman Empire. The construction thus began of a railway between Cernavodă and Constanța, which was inaugurated in 1860. The Dobrogea route, first thought of as a crisis solution to avoid the shifting sands in the Russian Danube Delta, now took on its own identity and created new opportunities for regional development, often in competition with the Danube waterway. The integration of Dobrogea in the Romanian state after 1878 amply demonstrated the value of routes of communication between the Danube and the Black Sea, a project that had also been massively popularized by passengers on the Austrian steamboats. As for the canal between Cernavodă and Constanța, the idea reappeared periodically in subsequent decades, until it was finally constructed, at considerable human and material cost, in the communist period.

EUROPE ON THE HILL

The introduction of steamboat services on the Danube and the Black Sea placed the Moldavian port of Galați on an important communication route between East and West. More and more travelers thus came to visit the town. Those arriving from upstream, from Austria, would stop in Galați for one or two days in the interval between disembarking from the river steamboat and embarking on the sea-going vessel that would take them to Istanbul or Odesa. Passengers coming from the Ottoman Empire and continuing their journey to Central Europe, on the other hand, were not allowed to enter the

41 Théophile-Sébastien Lavallée, "Les villes du Bas Danube," *Revue d'Orient* 14 (1853), 403–404.
42 Anon., "The Danube Ship Canal, and a Free Port in the Black Sea," *Leeds Mercury* (November 10, 17, and 24, 1855).

town without first passing through quarantine procedures, with the result that more often than not they only "saw" Galați from a distance, from the lazaretto area where the steamers anchored or from the Dobrogean bank (Figures 11 and 12).

Travelers recall the size and, above all, the dynamic character of the town, which was considered one of the most active ports in the Black Sea region. The Swiss Jacques (Jacob) Mislin (1807–1878), a Catholic prelate influential in imperial circles in Vienna, remarked on the principal traits of a town that was "commercial, dirty and bustling."[43] Both bustle and dirt were commented on by numerous steamboat passengers. Among them was Ida Laura Pfeiffer, who arrived in early April of 1842 at Galați, "the place of rendezvous for merchants and travelers from two quarters of the globe, Europe and Asia. It is the point of junction of three great empires—Austria, Russia, and Turkey." Pfeiffer spent a day in the town, walking "up hill and down dale through the ill-paved streets" and observing the cosmopolitan character of the place.[44]

Most visitors commented on the modernizing virtues of trade, visible in the very structure of the town. Galați was made up of two parts, clearly delimited both geographically and administratively: one part in the valley (the "lower town," situated at the level of the river) and the other uphill. Passengers disembarked in the lower town, which had a typically "oriental" appearance, with narrow, dirty streets. By comparison, the upper town, built from the proceeds of the growing cereal trade, seemed modern, civilized, "European." The 1837 Murray's guide mentions the recent development of the town and quotes a description from the previous year by Saint-Marc Girardin. The French man of letters remarks on the confused mixture of rickety houses and irregular streets paved with wooden beams and covered, depending on the season, in dust or mud:

> Picture to yourself, upon an eminence sloping rapidly to the waterside, a confused cluster of wooden huts, intersected by irregular streets, paved with trunks of trees placed from one side to the other; when it is fine weather, a tremendous dust converted by rain into deep mud. All manner

43 Jacques Mislin, *Les Saints Lieux. Pèlerinage à Jérusalem*, vol. 1 (Paris, 1851), 65.
44 Pfeiffer, *Visit*, 34–35.

Figures 11 and 12 Ludwig Ermini and Alois von Saar, *Galați* (*c.*1824).

of unwholesome smells issue from the stagnant pools which at all times collect beneath the logs;—imagine these cabins, dark and sombre within, and without filthy with mud, a sorry caravansera by way of inn, with apartments almost without furniture, and as full of dust as the streets; not the least appearance of any order, cleanliness, or arrangement; a town constructed like an encampment, and such an encampment as French soldiers would not put up with for a week together;—such is Gallatz, that is to say, Old Gallatz, the Turkish town, the aspect of which made upon me the same unfavourable impression that other Turkish towns on the Danube had done. At a distance, the mixture of habitations and verdure seemed inviting and graceful—the view of the interior destroyed the delusion. Fortunately, by the side of Old Turkish Gallatz a new town is rising, which will date its origin, like Brahilof [Brăila], from the regeneration of the Principalities. Upon the hill overlooking the Danube, a few buildings have already sprung up bearing a European aspect, and giving promise of what Gallatz is likely to be in future.[45]

Subsequent editions of the guidebook kept Girardin's description, updating the information about the lazaretto and advising travelers to be adequately prepared to face the attacks of mosquitoes and malarial fever.[46] Most descriptions are similarly constructed, with travelers insisting on the contrast between the two districts and on the vertical transition from the Orient to the beginnings of the West, from the bustle and squalor of the Levant to the order and aesthetics of Europe. Charles William Vane, marquess of Londonderry, describes the town, which he visited in late October 1840, in the same terms as Murray's guidebook:

The old town, bordering on a marsh, and flanked by the sea, is a mass of irregular wooden houses; the ways—for streets there are none—are a cloud of dust in summer, and a quagmire in winter. [...] The new town of Galatz is building on a height above the old one. The houses promise to be much better. Here all the consuls reside, as well as the governor and the commandant of the place.[47]

45 *A Handbook* (1837), 391. The French original in Girardin, *Souvenirs*, 240–241.
46 *A Handbook* (1858), 537.
47 Vane, *A Steam Voyage*, 147–148.

His wife Frances Anne Vane remarks along similar lines on the unpaved streets and "mere mud huts" of the old town, but also observes the large number of ox-drawn carts carrying grain to the harbor and notes that "the trade and commerce are increasing, and a new town is rising on a height, where the consuls have their houses."[48]

Another description of around the same period presents the lower town, "built very much in the style of European Turkey," which "from a distance, bears a pretty rural character, mingled as are the numerous trees with the red roofs of the low houses," though on a closer examination the author notes the filth, narrow lanes, and rickety houses. The upper part of Galați, on the other hand, "contains houses built with greater solidity." The population is increasing "and the importance of the place becoming daily more obvious, since the establishment of the Danube steamers," and there are hopes that, with wise investment on the part of the Moldavian ruler Mihail Sturdza, Galați will soon "bear something of a European appearance."[49] The Frenchman Xavier Marmier, too, contrasts the lower part of the town, "along the river" with its "old houses, poorly built, warehouses, shops," with the "new town, constructed with more taste and elegance." From up there, the view is superb, showing also the aesthetic virtues of commercial prosperity, most clearly visible in the harbor crowded with Greek, Turkish, and Italian vessels.[50]

The most visited part of the town had few attractions. Mud appeared to be ubiquitous in the region.[51] It seemed to have impregnated the very air and affected the entire environment, leaving its imprint on "the pale faces of its squalid and spiritless inhabitants, who crawled about their daily avocations with an air of listless indifference," a sure sign of their "moral degradation."[52]

The general appearance of the town in fact reflected the image of Moldavian society as a whole in a period of sustained institutional, administrative, economic, and cultural modernization. The authorities in Iași were themselves aware that the appearance of Galați could be generalized to the level of the whole Principality and that there was a need for more investment in public works in the country's most visible place. In 1849, they considered

---

48 Vane, *Narrative*, 90.
49 Anon., "The Mouths of the Danube, from the Notes of a Recent Traveller," *Colburn United Service Magazine: A Naval and Military Journal* 2 (1844), 199–200.
50 Marmier, *Du Rhin au Nil*, vol. 2, 14–16.
51 Skene, *Wayfaring Sketches*, 235, 240.
52 Anon., "The Danube," *Fraser's Magazine for Town and Country* 49, no. 293 (May 1854), 575.

that Galați ought to "present itself in the best conditions as regards its works and facilities, so that, according to its appearance, the powers that frequent its port should have about the whole country the opinion that it ought to inspire."[53]

Caught between the heritage of the medieval market town, which orbited around the Danube port, and the new district "on the hill," the measure of the prosperity of the entrepreneurs who were profiting from the penetration of Danubian products into the circuit of international trade, Galați was changing rapidly. It was clear to all tourists who arrived there that the Danube, the numerous commercial vessels in the harbor, the carts carrying grain, and the roads were all part of the vital transport infrastructure that fed the prosperity of the town and of the whole Principality of Moldavia. Agriculture was the staple occupation in the Principalities, and their ports owed their development to the period of free trade after 1829.

Russophobe contemporaries were convinced that Russia was not looking calmly on this prosperity. The foreign merchants in Galați noted that the ports of the Principalities and those of Russia on the Danube and the Black Sea (Reni, Izmail, and, above all, Odesa) were dealing with more or less the same products. The belief that Russia was aiming to sabotage the flourishing of Brăila and Galați became general after the establishment of the quarantine cordon on the Sulina branch. Passengers on board the steamboats took up the theme. In 1839, the British naval officer Adolphus Slade described the exports of the two Principalities as "various and most abundant, particularly in corn, wool, and fruits," expressing his opinion that "Southern Russia begins to feel the competition of Moldavia and Wallachia, and I doubt not that in a few years Odessa and Taganrok will decline in consequence."[54] In 1852 Oliphant noted that the Principalities were "annually becoming more formidable as rivals" to the Ukrainian provinces of the Russian Empire.[55]

The rivalry between the Danube ports and Odesa was a common theme in travel literature, and one by means of which the authors tried to impose a logical frame on Russia's policies regarding navigation through the Danube mouths. Russian "indifference" had to have a deeper motivation, in line with

---

53 Cezar Bejan et al. (eds.), *Tezaur documentar gălățean* (București, 1988), 114.
54 Slade, *Travels*, 200.
55 Oliphant, *The Russian Shores*, 345.

the image of the autocratic empire as perceived among the Russophobe public of the day.

"One of the Most Fertile Countries of Europe"

The external trade of the two Principalities was dominated by cereal exports. The ever-greater quantities of grain available for the export market resulted from the enlargement of the cultivated surfaces and from better management of agricultural production, against the background of the integration of the Romanian lands in the global capitalist system.

Agriculture features prominently in foreign travelers' descriptions of the Principalities. Those who actually visited Wallachia and Moldavia have left interesting accounts of the structure of landholding, production practices, agricultural equipment, social relations, and so on. In all these respects, the principal characteristic is backwardness; the cultivation of the land in the region was as rudimentary as all other human activities. Indeed, signs of progress could be seen, but the fact that the Principalities exported grain had a simpler explanation: the fertility of the soil coupled with the relatively low population density.

Today, too, the first part of this explanation is recalled in various forms whenever there is an attempt to explain the failings of Romanian agriculture, an endless "lost bet" throughout the country's recent history. One cliché is that Romania can easily supply the food needs of its population, and indeed even more. "We could have fed all of Germany, but we feed just a quarter of Romania," announced the title of an article in an economic publication in 2009.[56] "Romania can produce food for 80 million inhabitants, but its agriculture is subsistence-level," stated a Romanian Member of the European Parliament a decade later.[57] Various internal and external factors are invoked to explain the missed opportunity.

A historical analysis shows how long the themes of the exceptional fertility of the soil and the incapacity of the locals to profit from this precious

---

56 Anon., "România după 20 de ani: Am fi putut hrăni toată Germania, dăm de mâncare la doar un sfert de Românie," *Capital* (December 21, 2009), www.capital.ro/romania-dupa-20-de-ani-am-fi-putut-hrani-toata-germania-dam-de-mancare-la-doar-un-sfert-de-r.html (accessed August 18, 2023).

57 Anon., "România poate să producă hrană pentru 80 de milioane de oameni, dar agricultura e de subzistență," *Newsweek România* (April 8, 2019), https://newsweek.ro/economie/romania-poate-sa-produca-hrana-pentru-80-de-milioane-de-oameni-dar-agricultura-e-de-subzistenta (accessed August 18, 2023).

natural resource of the Romanian space have been around. The travel literature of the nineteenth century is full of relevant allusions. From all sorts of empirical observations—the diversity of crops to the size of the thistles on uncultivated fields—let it be understood that the Principalities were fruitful lands. The renown of Moldavia and Wallachia as "granaries" of Istanbul, and the extent of the land that lay uncultivated, contributed to the formation of this opinion. Writers who knew the country at close quarters confirmed the exceptional fertility of the Danubian plains. William Wilkinson (?–1836), British consul in Bucharest from 1813 to 1816, noted that "the fertility of the soil is such as to procure nourishment for ten times the number of the present population, and leave wherewith to supply other countries." However, in the absence of "the important advantages of a regular government and a wise administration, under which industry and agriculture should receive their due encouragement," this potential was far from being realized.[58]

Passengers on the Austrian steamers also noted various aspects concerning the natural productivity of the soil north of the Danube and charged it with various political significances.

The freedom of trade after 1829 and the growing demand for cereals on the European markets gave rise to a veritable revolution in the agriculture of the Danubian region, leading to an exponential growth in the areas cultivated and in production for commerce. In spite of the more intense use of agricultural land, however, the Principalities were insufficiently exploited in comparison to the great economic opportunities that presented themselves. On the way between Smârda and Giurgiu, writes Frances Anne Vane, "there was no road, no cultivation: on all sides we beheld utter waste and misery."[59] These unused lands were the sign of decline and lack of entrepreneurial qualities among a subjugated people. The population of Wallachia, Xavier Marmier estimated, was no more than 1.2 million, but the country could feed six times that number. "It is a shame to see such good land abandoned, while in other countries the needy peasant clears and cultivates the smallest piece of land with such care."[60]

58 William Wilkinson, *An Account of the Principalities of Wallachia and Moldavia with Various Political Observations Relating to Them* (London, 1820), 84–85.
59 Vane, *Narrative*, 82–83.
60 Marmier, *Du Rhin au Nil*, vol. 2, 5–6.

For Girardin, this "miraculous fertility" had saved the Principalities from many of the abuses of the Ottoman period.[61] In the new context, it ought to bring them not only material prosperity but also complete political liberation. After they had resisted for centuries, it was time for this resource to be used to obtain their independence from the Russians. Slade similarly noted, "During three centuries the history of the provinces has been a sad record of oppression: their exuberant fertility, defying tyranny, has alone saved them from ruin." Their products—cereals, wool, and fruit, but also wood and livestock—were sources of great wealth.[62] However, there was a need for more entrepreneurial culture in order for the benefits of this abundance to be felt in society. Without sufficient commercial exchanges, in Oliphant's view, the situation of the Principalities was badly affected. With more trade, he argues, Moldavia could become "one of the richest, as it is one of the most fertile countries in Europe": "Intersected by noble rivers, it only requires the properly directed skill of the engineer to render them navigable; blessed with a most magnificent soil, it only awaits the operations of some enterprising capitalist to yield its abundance."[63]

The writers also offered all sorts of reasons to explain the relatively difficult state of agricultural production in the Principalities: their underpopulation, the oppressive social system that did not encourage work, or the proverbial laziness of the local population. Agriculture thus became another factor in the Orientalizing of the Romanian space, a space well endowed by nature but incapable of profiting from the resources it possessed. What was lacking was the entrepreneurial culture and the mercantile disposition, which were seen as specific to the West and necessary if an opportunity was to be transformed into a reality.

EARTH-HOUSES AND DEGRADATION

As beautiful and fertile as the Principalities were, so were their inhabitants barbarous and degenerate. The degeneracy of the population was total, physically and morally, providing a demonstration of the direct effects of political dependency on the human body.

---

61 Girardin, *Souvenirs*, 237.
62 Slade, *Travels*, 197, 200.
63 Oliphant, *The Russian Shores*, 344.

The material backwardness of the local population was visible in two aspects mentioned by numerous passengers: their dwellings and their favorite food. The partly underground earth-houses (Romanian: *bordeie*) that could be seen in some riverside villages were symbolic of the sunken physical and moral state of their inhabitants.[64] Seen from the steamboat, a village looked more like a cluster of molehills or a termites' nest. Traveling downstream the river in 1843, German botanist Karl Heinrich Emil Koch (1809–1879), professor at the University of Jena, made reference in such terms to the local inhabitants living half-buried in squalid conditions.[65] Mislin similarly compared the peasants to moles living in squalid earth-houses and feeding on *mămăligă* (maize polenta).[66] That not only foreigners observed these realities is clear from a text by the Transylvanian Romanian historian and journalist George Bariț (1812–1893). Traveling to Giurgiu in 1836, Bariț could not hide his amazement at the sight of villages of earth-houses "one after the other, that had neither cowshed nor barn nor henhouse." And he continues: "You are moved to tears when you pass by those dens in which only wild animals should live and you see land beyond what the eye can see unworked, possessed by no one, so that if somehow, as they say, Wallachia has only a million inhabitants, going by how much land lies empty and unworked, with good management, another three million could easily live."[67]

The physical degradation of the peasants was also the result of their diet, and the *mămăligă* that led to atrophy of the body was often mentioned by travelers.[68] The men's decline was clear from their prematurely aged bodies. They were "well built, active and often naturally clever," but "idleness and the results of oppression" had deformed their natural vigor. Consumption of alcohol also contributed to their degraded state.[69]

---

64 On peasant dwellings and discursive attitude to *bordeie*, see Constantin Bărbulescu, *Physicians, Peasants and Modern Medicine*, trans. Angela Jianu (Budapest, 2018), 66–89.

65 Karl Heinrich Emil Koch, *Wanderungen im Oriente, während der Jahre 1843 und 1844: Reise längs der Donau nach Konstantinopel und nach Trebisond*, vol. I (Weimar, 1846), 100.

66 Mislin, *Les Saints Lieux*, 53.

67 *Călători români pașoptiști*, ed. Dan Berindei (București, 1989), 76–77; Violeta-Anca Epure, "Aspecte de viață cotidiană în principatele române prepașoptiste surprinse de consulii și voiajorii francezi: așezările, casele, arhitectura," *Cercetări Istorice* 37 (2018), 271–288.

68 On *mămăligă* see the recent book by Alex Drace-Francis, *The Making of Mămăligă: Transimperial Recipes for a Romanian National Dish* (Budapest, 2022).

69 Warrington Wilkinson Smyth, *A Year with the Turks: Or, Sketches of Travel in the European and Asiatic Dominions of the Sultan* (New York, 1854), 21–22.

The women were superior to their husbands. They are often described as having a pleasing appearance, "with expressive countenances and elegant forms, straight and supple," as Slade puts it.[70] In domestic matters, the women were presented as being very hardworking and capable of maintaining large families.[71]

The social structure had the same flaws. The population was subject to a form of personal dependence in relation to local governors (Romanian: *ispravnici*) who had the power of veritable masters. The common people were "virtually slaves," writes Elliott, emphasizing the "system of tyranny" that he found generalized in the Principalities. Violence was prevalent, divorce common, and the system of justice was primitive. "It may safely be affirmed that Christendom does not contain a country more demoralized and more degraded than Wallachia and Moldavia," he remarks. The rich were "given up to display, indolence, and political chicanery," while the poor were "in a state of abject misery and degradation."[72] Positive aspects were seen only when a comparison was made with populations considered even more backward. Army captain James John Best compares the Moldavian peasants with their Bulgarian counterparts and finds the former to be better off. They "were all well-dressed, and looked happy and clean," he remarks, and suggests that this may be attributed to "their lately acquired independence of the Turkish yoke."[73]

Though they might be "relapsed into an almost primitive barbarism,"[74] the Romanians were a noble people. Their present sordid state was the result of the loss of their connection with civilization. If one looked attentively beyond the layer of squalor, there was something noble about the inhabitants of Wallachia and Moldavia, which recalled their ancestors and their ancient origins. By their name and their language, monuments to Roman antiquity, the Romanians deserved to shake off the chains of dependence and slavery. Thus, the inhabitants resembled the country: an interesting mixture of Latin elements with Russian, Oriental, and European borrowings.

---

70 Slade, *Travels*, 169.
71 Epure, "Imaginea femeii din Principatele Române în perioada prepașoptistă în viziunea consulilor și călătorilor francezi," *Terra Sebus. Acta Musei Sabesiensis* 5 (2013), 403–416.
72 Elliott, *Travels*, 159–163, 178.
73 Best, *Excursions*, 297.
74 Smyth, *A Year*, 18.

Conclusions

A border is always a special place. Often it is a mere symbolic marker, a line drawn on a map that separates distinct political entities. In the case of a great international river that also played the role of an interimperial frontier, the navigable character of the Danube offered passengers on board the steamboats a privileged viewpoint from which to observe what was specific on either side. The Bulgarian shore was more picturesque and interesting, but the Wallachian and Moldavian side was attractive for its very ambiguity, as a space not only of separation but also of transition between empires and civilizations.

The geographical imaginary[75] is constructed in an ad hoc manner, on the basis of fluid mental maps. Direct observation, discussions with other travelers, short visits to the major Danube ports, these were the principal ways in which steamboat passengers became familiarized with Romanian realities. Documentation from written sources, especially guidebooks and travel narratives, already offered tourists basic information about the region, albeit often contaminated by the Orientalist and Russophobic discourses specific to the period. Thus, travel guides became part of a network of knowledge and power, fixing the geography of the region and populating it with a variety of other "realities."

In their accounts of their journeys, the travelers thus present an approximate, fragmentary image of the territory they crossed, made up of generalities, clichés, and prejudices. If the essence of the Ottoman, Oriental component of the local identity centered around the poverty and decline in which the Principalities and the local inhabitants themselves had lain for some centuries, much more complex is the image of the forms of Russian control over the region. The Russian Empire had contributed to the lifting of Ottoman monopolies regarding trade and navigation, but the foreign writers considered this a mere bait to entice the Principalities fully into the tsarist net. The power of Russia was based on covert forms of control, which were soon to show themselves openly, breaking through the veneer of apparent reformism.

Russia had, however, contributed to the release of latent entrepreneurial energies. Free trade, seen in a rationalist spirit as intrinsically bound to

---

75   Derek Gregory, *Geographical Imaginations* (Cambridge, MA, 1994).

civilization, had the capacity to modernize Romanian society and to free it completely from the past. In the main ports of the Principalities, Brăila and Galaţi, the progress was clearly visible, and "Europe" was gradually making a place for itself in the form of material prosperity, which tended to push aside both Oriental decadence and Russian corruption.

Thus, seen from the Danube, the Principalities were a fascinating contact zone between several empires. The superposition of two barbarous forms of domination, Turkish and Russian, contributed to the creation of an interesting political hybrid. The Principalities were an intermediary space, characterized by fluidity and uncertainty, but also by openness and opportunity. Through exchanges with "Europe," it was possible for "civilization" to be implanted there too, after having disappeared in the dark centuries of history.

Geographical and historical aspects confirmed these expectations. The north Danubian space was a fertile territory that could feed a numerous population. Materially and morally degraded as they were, the local people nevertheless had multiple qualities that could be put to use under a more just government inspired by Western models of good practice.

While, due above all to the connections offered by the Austrian steamboats, the "Europeans" were discovering the Romanian lands, at the same time, thanks to the same easy links, the Romanians were discovering "Europe."

CHAPTER 4

The Romanians in Europe

A MOLDAVIAN ADVENTURER

One of the most fascinating characters in the Irishman Michael J. Quin's account of his journey on the Danube—a book that enjoyed considerable success in the mid-1830s—is the "Moldavian adventurer," already mentioned in Chapter 2. His appearance immediately makes him stand out on board the Austrian steamboat. He wore a tattered blue frock coat, "a pair of old black stuff trousers patched at the knees in a most unworkmanlike manner [...], together with a ghost of a black waistcoat, a cast-off military cap, and wretched boots." "He had not shaved for three weeks—he certainly could not have washed either his hands or his face for three months, and a comb had probably not passed through his hair for three years," continues Quin's sarcastic portrait of the man, completing the portrait with "a very red nose, on the top of which was perched a pair of spectacles."[1]

Despite his less-than-attractive appearance, the adventurer was popular with many of his fellow travelers. Third-class passengers—"a miscellaneous group of Austrian soldiers and their wives, pedlers, and artisans, who occupied mats and sheepskins on deck"—seem to have particularly taken to him, and so did the crew. "He whistled well, he sung well, and passed off every thing in a 'devil-may-care' kind of way, which gained him admirers," Quin remarks. He had a commonplace book from which he sometimes recited "scraps of poetry of his own composition, or selected from the works of celebrated German writers," adding "comments often so droll, that he set the

---

1   Quin, *A Steam Voyage*, 17–18.

whole deck in a roar." On other occasions, he would tell of his own travels and adventures, in such diverse places as Bucharest and Istanbul, Prague and Vienna, St. Petersburg, Paris, Berlin, Madrid, Gibraltar, and Venice—"every where but London, where he had the modesty to confess he had never yet been."[2]

Quin was intrigued by this figure with ragged clothes but "the flush of fine intelligence" in his face and who, among his numerous qualities, spoke German, French, and Italian fluently. He had taken part in the Russian–Turkish war of 1828–1829 and might very well be a Russian spy.[3] The Irishman and the Moldavian became closer as the number of passengers was reduced, with only those few who were traveling downstream of the Iron Gates remaining on board. In the Iron Gates region, the Moldavian adventurer's abilities helped make the continuing journey more pleasant. The passengers spent the night in the village of Sviniţa, where he displayed his multiple talents in the course of an impromptu party in the local inn. After recounting more of his Oriental adventures, our hero vanquished the village priest in a theological debate and then offered a rendering of Rossini's *Di Tanti Palpiti* that astonished all present, travelers and locals alike. He sang, Quin remarks, "not only with great taste, but in one of the best tenor voices I ever heard." The villagers too were amazed, and "the priest exclaimed that he knew not what to think of this fellow, unless he was the devil, for that not only were his talents and knowledge universal, but of a degree of excellence in every thing that left him without a rival."[4]

Indeed, Quin himself confesses that, when he considered the Moldavian's multiple abilities and fields of knowledge, he "could not help feeling that there was a mystery about him, such as perhaps in a former age might have procured for him the dangerous honours of a magician."[5]

The British historian E. D. Tappe identified the adventurer as *sluger* Teodor (Tudorache) Burada (1800–1866), a figure little known in Romanian historiography. The son of a priest, Burada studied theology at the seminary in Iaşi, where he acquired a solid musical education. Suspected of being implicated in a plot against the Moldavian ruling prince Ioniţa Sandu

---

2   Ibid., 18–19.
3   Ibid., 17, 20.
4   Ibid., 98–101.
5   Ibid., 102–103.

Sturdza (1760?–1842, reigned 1822–1828), he fled to Wallachia, where he was employed as a music teacher in Craiova and Cerneți and then traveled through Europe. Returning to Moldavia, Burada held various public positions and was an active supporter of local cultural life. Almost all the information we have about him supports the hypothesis that Burada was Quin's "adventurer," one of the first Romanians we know to have traveled by steamboat on the Danube.[6]

The steamers of the DDSG accelerated mobility in southeastern Europe, including in the Principalities of Wallachia and Moldavia. As I demonstrated in the preceding chapters, the great river provided the easiest connection of the two states with the wider world, and the Austrian steamboats favored the integration of the Romanian space in "modernity." Just as more and more foreigners ventured into the relatively unknown territory of the Principalities, so the political and economic elites of the Romanian space were quick to embrace the new means of transport in order to travel both to the West and to the East.[7] In this chapter, I shall examine this aspect, with a special focus on the way in which Romanian writers perceived the steamboat and travel by steamboat, the experience of Romanians on board the vessels, the cost of voyages, and the role of the steamers as "revolutionary" machinery.

THE SPECTACLE OF MODERNITY

The Austrian steamboat was the principal means of transport that connected the Principalities with the wider world, the vehicle on which locals embarked for destinations in the "civilized" West or the "exotic" Orient. The steamboat was the first motor of the industrial revolution that the Romanians encountered at first hand in their own homeland.

After 1834, steamers became a familiar presence in the ports of the Lower Danube. Their specific impact on the senses made their presence immediately obvious: the trail of smoke rising from the funnel, the rumble of the

---

6 E. D. Tappe, "Was Quin's 'Moldavian Adventurer' Slugerul Burada?" *Revue des Etudes Sud-Est Européennes* 12, no. 4 (1974), 588–590.

7 For details about Romanians' travels and travel memoirs, see, among others: Florin Faifer, *Semnele lui Hermes. Memorialistica de călătorie (până la 1900) între real și imaginar* (București, 1993); Constantin I. Stan and Alexandru Gaiță, *Călătorii ale românilor în centrul și vestul Europei (1800–1848)* (Buzău, 2013); Mircea Anghelescu, *Lâna de aur. Călătorii și călătoriile în literatura română* (București, 2015).

paddles, the smell of burning coal—all individualized the steamboat in a period when such vessels were still rare on the waters of the Danube. Curious locals would stop to admire the technological wonder passing along the river or would go watch the passengers disembarking at the ports. "Wherever the boat stopped," records Julia Pardoe (1804/6–1862), "crowds of the peasantry flocked to the edge of the water, and stood gazing at her in admiring wonder; for, as this was only her twelfth voyage, their curiosity and astonishment had not yet subsided."[8] Charles B. Elliott writes that the arrival of a steamboat was a veritable "gala-day": "On these occasions the vessel becomes a general rendezvous for all the gossips of the place, and ordinary recreations and amusements are absorbed in that superlatively gratifying one, seeing and being seen, talking and being talked to."[9] In other Danubian territories too, the situation was similar. The arrival of the steamer at Zemun was an event that seemed to capture the attention of the whole town. George Robert Gleig remarks how the pier was "crowded with spectators," some hoping for news of "what was going on at Pesth? how stood matters at Vienna?"[10] When the steamboat on which the physician John Mason (?–?) was traveling arrived at Tulcea in 1846, the "motley population crowded to the landing-place to receive us."[11] Part of the usual bustle of the harbor at the moment when the steamer departed with its human "cargo" consisted of friends and relatives of the passengers, there "to embrace those who were leaving and parting from them, and these, for their part, were addressing greetings and farewells to all who then surrounded them."[12]

As there were almost always important people on board the steamboats, the authorities gave particular attention to their new connection with the world. Pardoe was on board the *Ferdinand I* during a princely visit to the port of Galați. She describes how first the princess of Moldavia was honored by the captain with a thirteen-gun salute as she passed, while the following day the prince himself, Mihail Sturdza, announced by the fifes and drums of the local garrison, was welcomed on board the steamer.[13] Some years later, on October 15, 1843, the arrival of the prince of Wallachia, Gheorghe

8   Julia Pardoe, *The City of the Sultan: And Domestic Manners of the Turks in 1836*, vol. 2 (London, 1837), 428.
9   Elliott, *Travels*, 205–206.
10 Gleig, *Germany*, 274–275.
11 John Mason, *Three Years in Turkey: The Journal of a Medical Mission to the Jews* (London, 1860), 44.
12 Pelimon, *Impresiuni*, 137.
13 Pardoe, *The City*, vol. 2, 420–424.

Bibescu, at Smârda (Giurgiu), on his return from his investiture at Istanbul, was also a noisy one:

> The steamboat Argos discharged a gun to announce the arrival of its illustrious guest. At once the guns of the quarantine station began to fire continuously. When the steamboat had come close to Smârda, it discharged another two guns and drew up to the specially decorated pier. His Highness came out of the steamboat together with all his suite, and, to the sound of guns, was met [by an impressive reception committee of dignitaries and a crowd of common people].[14]

Various institutions on the banks of the Danube were duty-bound to offer distinguished travelers a fitting salute. In 1847, the authorities in Brăila recorded the purchase of 15 *ocas* (about 20 kg) of gunpowder for "the salute made on the part of the quarantine station on the occasion of the arrival of His Highness the Prince and the Russian steamboat" plying from Odesa to the Danube; other expenses were incurred by the Gunboat Service (*Servicul șeicilor canoniere*), the "border police" of Wallachia, with "the filling of the guns discharged on the passage by steamboat" of the Russian consul general Dashkov.[15]

Between such audible manifestations and the other sensory elements mentioned above, the passage of a steamer and the presence of its notable guests were hard to ignore.

A CONNECTION TO THE WORLD

As the fastest and simplest means of communication with the world, the Austrian steamboat was the courier that transported back and forward between East and West not only travelers and goods but also information. Passengers carried with them news of epidemics, political crises, and social disturbances, which they spread all along the route. As the first telegraph lines in the Principalities were not installed until the period of the Crimean War, such news, provided in written reports or spread informally by word

---

14   Isar, *Sub semnul.*
15   *Analele Parlamentare ale României*, vol. 16, part 1, *Divanul Obștesc al Țărei Românești, Legislatura V, Sesiunea I (XV), 1850–1851* (București, 1909), 337, 469.

of mouth, could be crucial for the security and prosperity of the people of the two states. At Giurgiu, Hans Christian Andersen recounts, a number of townsfolk were in the vicinity of the port to ask "about the state of health in Constantinople, and about the disturbances in the country"[16] (the reference being to a revolt of the Bulgarians in the Ottoman Empire). In 1850, the news of clashes between Muslims and Christians in Aleppo were brought to Moldavia by travelers arriving on the Austrian steamer.[17]

As mentioned in Chapter 1, in 1846, the steamers transported approximately six thousand travelers on thirty-six sailings between Istanbul and Galați/Brăila (eighteen in each direction), together with 4,430,073 florins in secure bags, 18,635 letters, 16,291 parcels, and 536 packets.[18] It is superfluous to insist on how important all these "goods" were for the business of merchants and for the integration of the products of the Principalities in the international circuit.

The quantity of cash in secure bags transported by the Austrian steamboat, immediately after the connection between Galați and Istanbul became operational in 1836, caused currency speculation to flourish in the Moldavian port. In May 1837, the government in Iași decided to verify these coins, in order to reduce speculations that were affecting the exchange value of the local currency.[19]

That "knowledge" too circulated faster along the steamboat route is clear from the example of Mihail Kogălniceanu. While studying in Berlin, he asked his father for a number of volumes he needed in his work on his doctoral thesis. The books could be sent by Danube steamer, just as the studious young man would later send a large part of his library "on the Danube to Galați by the steam vessel."[20]

The circulation of correspondence also became faster and more punctual. The steamboat was the courier that ensured connection to the world, enabling those traveling for reasons of pleasure or necessity to keep in contact with their homeland. In October 1848, Teodor Râșcanu wrote from Istanbul to his brother Alexandru, to whom he had sent a letter through

---

16 Andersen, *A Poet's Bazaar*, 126–127.
17 BAPR, vol. 1, part 1, 253 (*Jurnalul de Galați* 2, 1850).
18 *The Overland Mail*, 8–10.
19 Cristian-Dragoș Căldăraru, "Orașul Galați în documentele din Manualul administrativ al Moldovei, 1834–1852," *Danubius* 32 (2014), 172.
20 M. Kogălniceanu, *Scrisori, 1834–1849*, ed. Petre V. Haneș (București, 1913), 73, 107, 133, 180.

Grigore Arghiropol (1825–1892), "who is coming to Moldavia for his busi-ness. [...] I am happy that with the coming of the previous steamboat a letter came to me from Father Iosif in which he tells me that you are all healthy." Later, on April 26, 1856, Nicolae Istrate wrote to Râșcanu in Istanbul that he was awaiting "either a steamboat from Tsarigrad [Istanbul] or a post, with the most tense impatience and you may imagine yourself our desper-ation when we never find a letter from you."[21] On his way to Tulcea in July 1850, Ion Ionescu de la Brad (1818–1891) sent a message to Ion Ghica that he expected Ghica's letter "on the next steamboat" and that he would write more, also "on the next steamboat."[22]

The steamers' schedule regulated the circulation of information in the region, determining the operation of other services in the Principalities, such as the post and diligences. The steamboat also contributed to a new relation to time, standardized, synchronized, and measured with mechan-ical instruments in "clock societies."[23] We may take as an example Teodor Codrescu (1819–1894), who, while sailing to Istanbul in 1839, kept a care-ful note of the times at which the vessel arrived in various ports. Codrescu's exactness is impressive, indicating a habit or even mania of consulting his watch. The travelers left Galați on Saturday morning at 8:30. At ten minutes past eleven they were at Isaccea, at thirty minutes past twelve at Tulcea, and at five minutes past six in the evening at Sulina. The next day, they reached Cape Kaliakra at four minutes to nine, and at eight minutes to twelve they were in the Gulf of Varna, where they stopped for thirty-eight minutes. The lighthouses of the Bosphorus came in view on Monday at twenty-five min-utes past midnight, but it was only at ten minutes past two that Codrescu admired them "in ecstasy." Finally, at ten minutes past four, they arrived at Galata.[24]

---

21 Gh. Ghibănescu (ed.), *Surete și izvoade (Documente slavo-române)*, vol. 10, *Documente cu privire la fa-milia Râșcanu* (Iași, 1915), 380, 442.

22 Victor Slăvescu (ed.), *Corespondența între Ion Ionescu dela Brad și Ion Ghica 1846–1874* (București, 1943), 124.

23 Jürgen Osterhammel, *The Transformation of the World: A Global History of the Nineteenth Century*, trans. Patrick Camiller (Princeton, 2015), 71–76. References to the Romanian lands in Armin Heinen, "De-spre cultura tehnică a epocii moderne occidentale și perceperea cu totul diferită a timpului în România. Măsurarea timpului și timpul social din Evul Mediu până în prezent," *Analele Universității "Dunărea de Jos" Galați* 19, no. 7 (2008), 241–254.

24 Constantin Gane, *Domnița Alexandrina Ghica și contele d'Antraigues* (București, 1937), 137–141; Drace-Francis, *The Traditions of Invention*, 103–107.

ROUTES OF MOBILITY: THE DANUBE VOYAGE AND
THE BUKOVINA ROUTE

The steamboat opened up new opportunities of mobility for the Moldavians and Wallachians, who had already begun to roam the continent for business or pleasure. Their travel options were determined by simple geographical factors. From Wallachia, the Austrian steamers were the easiest means of getting to both the West and the East. As I mentioned in Chapter 1, Giurgiu was the port of embarkation for travelers coming from Bucharest. From Giurgiu it was possible to reach either Vienna or Istanbul in five days, with the duration of the westward journey also depending on the period of quarantine on entering the Habsburg territories. From Moldavia, there were a number of routes toward the West, one of them being overland via Chernivtsi in Bukovina, from where one continued by mail coach or diligence and then by railway. Travelers from the south of Moldavia, on the other hand, generally opted for the Danube route when they traveled west. Journeys to the Orient were almost always on board the Austrian steamers, which could reach Istanbul from Galați in two or three days. Galați, Brăila, and Giurgiu were thus important nodes in the transport network.

For travelers from Moldavia, the choice between a river or an overland route depended on a number of factors, from the time of year (as the Danube was not navigable in the winter) to considerations of cost and comfort. As shown by the examples in this chapter, both variants were used by Moldavian travelers, who integrated them efficiently in their travel plans.

For passengers embarking in Lower Danubian ports, the Austrian steamer was more often than not the first link in a long chain of means of transport through which they came to know the world. As discussed at length in Chapter 2, it was here that they were socialized into the expectations of modern travel. The 1846–1847 voyage of the Moldavian boyar Iancu Prăjescu (1803–1894) on his own Romanian version of the Grand Tour is illustrative of the complexity of some of the routes followed. After leaving Iași on May 20, 1846, Prăjescu embarked at Galați a few days later for Istanbul on board the Austrian Lloyd steamboat. After visiting the city, he set out for Athens on the French steamboat *Mentor*, visiting Smyrna [Izmir] on the way. Eight days of quarantine at Piraeus and nine days spent in Athens were followed by more voyages, first to Corfu and then to Malta. Prăjescu then continued, on board the *Herculanum*, to Naples, where he arrived after a terrible storm "[such] that the waves passed over the deck of the steamboat." A month later, another steamboat took him to Civitavecchia, from where he continued by diligence to Rome. There followed journeys by diligence and train to Florence, Venice, Milan, and Genoa. The Mediterranean portion of Prăjescu's European tour ended at Marseille, where he arrived on board the French steamer *Charlemagne*, after another terrible storm. He spent the autumn months in France, Britain, the Netherlands, and Belgium, on an enviable tour that took him to Lyon, Paris, Rouen, Le Havre, London, Ostend, Antwerp, Utrecht, Amsterdam, the Hague, Rotterdam, Brussels, and back to Paris. In December 1846, Prăjescu took the train to Cologne and then continued by diligence to Hanover and again by train to Berlin, where he stayed for eleven days. After shorter stays in Leipzig and Dresden, he spent twelve days in Munich and a week in Vienna. On January 27, 1847, eight months after setting out on his long European tour, our boyar headed back to his homeland using the Austrian diligence service, via Lviv and Chernivtsi.[25]

Another Moldavian boyar (or indeed he may have been the same Prăjescu) traveled from May 11 to August 6, 1851, by an equally complex route. He left Iaşi for Berlin via Chernivtsi, Lviv, and Cracow. There followed a succession of stays in Cologne, Paris, Brussels, Calais, and London (where the Great Exhibition was on!) and then back via Brussels, Cologne, Berlin, Dresden, Prague, and Vienna. There the boyar embarked on the steamboat for Galaţi, and so back to Iaşi. The journey cost 34,000 lei, a considerable sum for the period.[26]

The distinguished Moldavian prince, economist, and statesman Nicolae Suţu (1798–1871) set out in May 1839 on a European tour accompanied by his wife Ecaterina and their children Eufrosina and Constantin. After spending a few days in Vienna, the tourists headed for Venice and Milan. They then went over the Simplon Pass to Geneva, where they attended the examinations of their first-born, Alexandru Suţu. Then, as Suţu records in his memoirs, "From

---

25 George Potra, "Statele Europei la 1846–1847, văzute de un boier moldovean," *Revista istorică română* 9 (1939), 207–245.

26 Potra, "Călătoria unui boier moldovean în Europa la mijlocul secolului al XIX-lea," *Revista istorică* 19, nos. 4–6 (1933), 126–139; Potra, *Călători români în ţări străine* (Bucureşti, 1939), 55–56. Also see Dan Râpă-Buicliu and Iulian Capsali, "Însemnări din 'Jurnalul de călătorie în Occident' al boierului moldovean Iancu Prăjescu," *Danubius* 27 (2009), 235–262.

Geneva, after a picturesque excursion to Schaffhouse, we left for Paris, where we left Alexandru, and we returned to Iași around the end of the year."[27] The prince's record of his travels is rich in references to the pros and cons of the Bukovina route. In what follows, I shall present the course of his journey with a number of observations regarding the route between Moldavia and Vienna.

On May 22, 1839, the travelers left the family estate in Iași county, hoping to reach the border before twilight. Torrential rain frustrated their plans, however, forcing them to stop overnight in Botoșani. They continued on their way the next morning. Outside the town, "the ground was so soaked with rain from the previous day that the ten horses stopped all at once in the course of the climb from Cucorani; neither the furious shouts nor the telegraph-like gestures of the postilions could stop them falling back, instead of advancing." With the help of the load-bearing cattle of some locals, they managed to climb the hill, and at 11 o'clock, they reached Mihăileni, at the border between Moldavia and Bukovina. In the quarantine pavilion, they were questioned by a physician about their state of health and about their luggage: "We had to tell him of how many coats, underclothes, and ornaments it was composed, if we had weapons, letters, fruit preserves, tableware, tobacco, or oranges." Next, they stopped at the customs house itself, where they paid customs duty for the tobacco a foreigner was permitted to bring into the empire. The family arrived at Chernivtsi that evening and spent the following day there, "planning our journey and making provisions." They left Chernivtsi on May 25, after hiring the services of a carter "who had 4 very good horses." They paid him the very acceptable price of 13 ducats to take them to Lviv, half what it would have cost to go by mail coach. After an exhausting journey, they arrived in Lviv on the morning of May 28. The beauty of the landscape compensated for the effort of proceeding along some poor sections of road, made more difficult by the rain. They set out for Vienna by mail coach on May 29, after preparing themselves with all the information necessary for the journey to be as predictable as possible with regard to costs:

> To pre-empt this inconvenience, one requests from the place where the horses are changed an itinerary, on which for each stopping place is

---

27 *Memoriile principelui Nicolae Suțu: mare logofăt al Moldovei, 1798–1871*, ed. Georgeta Penelea Filitti (București, 1997), 160.

indicated the price of the mail coach, the amount for horses and tips, the amount for the stableman and for greasing the wheels. Armed with this information, the traveler knows all the expenses in advance, and the only effort he has to make is that of undoing his pouch to pay without protest. This information is of great importance, especially when you travel at night. I was assured that following numerous complaints addressed to Vienna by travelers, the mail coach service has been put in better order so as to free them from delays and obstructions.

The journey lasted "5 and a half days and 4 nights, during which we did not truly stop except at Bielitz."[28] During the day, they continued without stopping, and at dusk, "we would stop at the first inn that we came to, to dine on a steak, after which we prepared for sleep." Sleeping in the coach was difficult at the best of times, but Suțu found the first night completely unbearable:

The next day, I was extremely tired, but exhaustion helped me to drop off. However, this sleep in which the limbs are numbed by the haphazard positions that the movement of the coach imposes on the body, this sleep interrupted by stops that you are forced to make, either to change the horses or to pay the numerous tolls for bridges and highways, has nothing restful about it. Not even a cup of water, either to drink or to wash my eyes or at least rinse my throat; no question of smoking or even drinking a cup of coffee. Thus, I count among my moments of happiness those few minutes during which we stopped between 7 and 8 in the morning to wash and take coffee.

The journey to Vienna took twelve days and proved exhausting.[29] By way of comparison, let us consider the details of Mihail Kogălniceanu's journey to Vienna in the spring of 1844, accompanying his sister Elencu, who was going there for an ophthalmological intervention. On April 15, the travelers left Iași, arriving at Botoșani in the evening. The next day they spent at Mătieni, and on April 17, they crossed the border at Mihăileni and arrived in the evening

---

28   Today the city of Bielsko-Biala, in Poland.
29   Nicolae Suțu, *Amintiri de călătorie, 1839–1847*, ed. Petruța Spânu, Gheorghe Macarie, and Dumitru Scorțanu (Iași, 2001), 17–43.

at Chernivtsi. After spending a day there, they set out again on April 19 by *Eilvagen* (diligence), passing through Sniatyn and Kolomea, "one-time towns of Moldavia." They left Lviv in the evening of April 23, also by diligence, in the company of a Mrs. Cihac and her daughter. From Lipnik (today Lipník and Bečvou, in the Czech Republic), they continued their journey by railway, arriving in Vienna on April 28 (May 10 new style), thirteen days after leaving Iași.[30]

Travel by the Bukovina route was difficult too, requiring some ten days on the road, at a rate of some 120 km per day. The situation gradually changed as the highway network of the Habsburg Empire was improved and railways were built. The time spent in diligences and mail coaches was greater than that on the steamboat, but the costs were lower, especially for families traveling. Another advantage concerned the greater freedom to establish the calendar of the journey, given the greater frequency of overland transport services. As already mentioned, travelers from Iași used both routes, taking into account the time of year, the financial resources available to them, and the degree of comfort they sought.

THE LEVANT AND ITS PILGRIMS

The Austrian steamers greatly facilitated relations between the Romanian space and destinations in the Levant. Situated just five days away from the capitals Iași and Bucharest, and less than three days from Brăila and Galați, Istanbul was an important hub for international mobility, a gateway to the Orient and also to the south of Europe. For political, economic, or religious reasons, the route was in considerable demand.[31] The principal customers were merchants, but other social and professional categories also took advantage of the opportunity to get quickly and safely to Istanbul and elsewhere. "Travel through foreign countries," wrote Gheorghe Asachi, "became important also for those less well-off, who sought health, light, or entertainment through travel."[32] Among these, we find various clergymen going on pilgrimages to Mount Athos or Jerusalem.[33]

One of the pilgrims was the hieromonk and scholar Chiriac of Secu Monastery, mentioned in the Introduction. Chiriac embarked on an

30 Kogălniceanu, *Opere*, vol. 1, 487–488.
31 On the attraction of the Orient in Anghelescu, *Lâna de aur*, 101–138.
32 *Călători români pașoptiști*, 6, n. 10.
33 Details in Faifer, *Semnele lui Hermes*, 68–89.

Austrian steamboat at Galați on May 13, 1840. Although it was more expensive than a sailing ship, the monk had sound arguments for preferring the modern vessel:

> By sailing ship you have to go for at least four days on the Danube and various other inconveniences, and even *pulling the ship along the tow path*, which is on the bank of the Danube, meaning that you harness yourself with a rope round your chest that is tied to the ship, for it to go faster! But the trouble with mosquitoes, who can tell, for in all the world I consider that they have no empire like the plain of the Danube with its reeds, which you cannot take in with your eyes. However, setting out by steamboat at 8 hours of the morning, a summer's day, at 6 [in the evening], I was in the Black Sea.

At Istanbul, Chiriac embarked on a sailing ship and, sailing past the island of Lemnos, reached the Holy Mountain, where he stayed for almost a year. From there, he left on July 1, 1841, on a pilgrimage to Jerusalem and had fresh adventures along the route by Izmir, Rhodes, Beirut, Cyprus, and Jaffa (Tel Aviv).[34]

A group of monks from Neamț Monastery in Moldavia set out for Istanbul, and from there to the Holy Mountain, in June 1841. The monk Maxim the *hagiu* (pilgrim) and two novices set out from the monastery, "in the monastery coach, with four horses and a coachman," and embarked at Galați for Istanbul. According to the passport issued by the authorities in Galați on June 26, 1841, the novices Iosâp and Sămion "[are] to embark on the sailing ship, and his reverence has embarked on the steamboat."[35] Later, around 1853, the monk Andronic Popovici (1820–1893) of Neamț Monastery also traveled to the Holy Mountain and took another voyage there in 1858.[36]

---

34 G. Giuglea, "Călătoriile călugărului Chiriac dela Mănăstirea Secul. Călătoria la Muntele Atos și Ierusalim," *Biserica Ortodoxă Română* 54, nos. 3-4 (1936), 159; also Zahariuc, "Călătoria ieromonahului Chiriac din Mănăstirea Secu la Muntele Athos (1840-1841)," *Analele Științifice ale Universității "Alexandru Ioan Cuza" din Iași, s.n. Istorie* 61 (2015), 249-264.

35 Zahariuc, "Din corespondența unui călugăr român la Muntele Athos în secolul al XIX-lea: Maxim hagiul," in Zahariuc (ed.), *Relațiile românilor cu Muntele Athos și cu alte Locuri Sfinte (sec. XIV–XX). In honorem Florin Marinescu* (Iași, 2017), 197-198.

36 Ieromonahul Andronic, *Călătoria la Muntele Athos (1858–1859)*, ed. Petronel Zahariuc (Iași, 2015); Zahariuc, "Sur le hiéromoine Andronic des monastères de Neamț et de Secu et sur son voyage au Mont Athos (1858–1859)," *Analele Științifice ale Universității "Alexandru Ioan Cuza" din Iași, s.n. Istorie* 62 (2016), 151-197; Zahariuc, "Despre același ieromonah Andronic de la Mănăstirile Neamț și Secu, însă despre o altă călătorie: la Ierusalim (1859)," in Liliana Rotaru (ed.), *Historia est magistra vitae. Civilizație, valori, paradigme, personalități. In honorem profesor Ion Eremia* (Chișinău, 2019), 234-243.

Such pilgrims were still few in this period, but their numbers would grow in the second half of the nineteenth century, when more and more clerics and laypeople took advantage of the increasing ease of travel to visit the holy places of Christianity. The statistical data included in Chapter 1 underline the large number of travelers embarking at Brăila and Galați for Istanbul, with the steamboats carrying hundreds of passengers on every sailing.

A Democratic Space

For many Romanians eager to know the world, the steamboat was the first foreign territory on which they set foot, and, as pointed out in the preceding chapters, the mobile space of the steamer was a setting for intense socializing. In the cabin and on the deck, the Romanians became part of a global community into which they seem to have integrated quickly.

On his way to Vienna, George Bariț went by diligence from Brașov to Budapest in September 1845. From the Hungarian capital, he took the steamboat, a means of transport that he considered "democratic." Gathering together "people from all walks of life" in one place enabled an easier communication among the travelers. According to Bariț, there was less shyness on the water.

Here, the most arrogant monied aristocrat, the most pedantic philosopher of pretense—and seriousness, the grandest lady comes in need and begins a sort of speech; one asks you for a light from your cigarette; another asks you the name of the citadel on the left or right bank of the Danube; some speculator wishes to know your homeland and in the course of conversation has to find out its articles of commerce, just to pass the time, for the three *caiute*[37] (indoor rooms) and two or three little cabins on the deck of the vessel do not have enough corners for all the hypochondriacs and all the misanthropists to be able to withdraw or for gatherings to form by class [...] There are many other select persons whom I do not have room to present here; suffice to say that conversation was all day lively and in many respects interesting; politicking, satirizing, whispering, and shouting flowed freely, exchanging places one with another.[38]

37 A borrowing into Romanian from the German *Kajüte* (cabin on a ship).
38 Qtd. in Mircea Popa, "George Bariț—călătorul," *Anuarul Institutului de Istorie "George Barițiu" din Cluj-Napoca. Series Historica* 42 (2003), 89–99.

The German civil servant Richard Kunisch (1828–1885) also noted the importance of the society on board the "floating house" that, for a few days or weeks, became "the traveler's world":

> Of course, you have your cabin, to which you can withdraw, but the cabin is so small that you can only sit, stand, or lie down in it. If, however, you leave it, contact with the other passengers is inevitable, especially during meals, which are taken communally, because for each class there is only a single servant. Staying mute for days on end in the midst of a society that is making conversation is not, at the end of the day, more uncomfortable than being forced to speak with antipathetic persons.[39]

Mobility and nobility went hand in hand, and the Romanians could boast of the high-society company in which they had traveled. The steamboat company too was proud of the distinguished guests who traveled on board, who were also invited to leave their signatures in a book of honor. According to Octavian Blewitt, writing in 1840, the "album or 'Fremden-buch'" of the *Ferdinand I* included the names of "Prince Cantacuzene, Mavrocordato, the Persian princes. the author of the *Kussilbash*, Miss Pardoe, Colonel Considine, &c."[40]

Captains gave such notable figures special attention and organized dinners in their honor and in honor of those who merited congratulations for various achievements. After concluding an important business deal in Istanbul in 1836, the Moldavian boyar Iorgu Hartulari was treated to a festive reception on the *Ferdinand I* on his way home. The table was set with "some 20 bottles of champagne and glasses. The captain of the steamboat comes with a large champagne glass, into which he pours a whole jug and the servants who attend at table pour for everyone." As related by Mrs. Hartulari:

> My husband and his nephew wonder what this means! He sees that the captain of the steamboat comes and addresses all the people, inviting them all to drink in honor of boyar Hartulari, because he has concluded such a big deal. My husband seeing this and being rather boastful by nature, says to the nephew who was sitting next to him to bring him a bag of *icosar*s, in which there were 200 silver *icosar*s, and after dinner he takes

39   Richard Kunisch, *Bukarest und Stambul* (Berlin, 1861), 215–216.
40   Peregrine, "The Danube," 1, 564. (The author of *The Kuzzilbash, a Tale of Khorasan* (1828) was James Baillie Fraser, who, as mentioned in Chapter 1, accompanied the Persian princes on their Danube journey.)

two of those vases that were there with fruit and empties them and puts
in all the *icosar*s from the bag and then he takes the glass of champagne
and rises to his feet and expresses thanks, first to the captain and then to
all those who were at table, in the languages that he knows: Greek, Moldavian, German, and Hungarian. Then they all rise to their feet and take
another glass and all shout: Hurray! Boyar Hartulari.[41]

On his way to the Orient in the spring of 1846, the boyar Prăjescu traveled together with Aloys II (1796–1858, prince of Liechtenstein from 1836
to 1858), Count William Albert de Montenuovo (son of Napoleon's widow,
Duchess Marie Louise, and General Adam Albert von Neipperg), and a
number of Hungarian nobles.[42] In recording information of this sort in his
account of his journey, the boyar could feel himself part of an international
noble community.

Foreigners, too, noticed the Romanian passengers, who became more
numerous and noisier on the segments of the voyage in the region of the
Lower Danube. Von Adelberg, the Austrian consul in Syria, who traveled
between Zemun and Galați in 1841, met on the steamboat the Prussian
consul in Iași, C. A. Kuch. A number of "respectable boyars" among their
traveling companions "unanimously expressed the most favorable opinions
about Kuch," regretting only that he did not have "a satisfactory income."[43]
The disembarking of the numerous groups of Wallachians at Giurgiu came
as a relief to the traveling companions of the British geologist Warrington
Wilkinson Smyth (1817–1890), who traveled down the Danube in 1852. The
Wallachians had not been very popular on board after a servant of one of the
Ghica princes had threatened another passenger and the prince had behaved
arrogantly. The captain and other passengers intervened on the side of the
abused passenger, and the culprit was kept in arrest until it was time for him
to disembark.[44] The Frenchman Jacques Boucher de Perthes remarked on his

41 Elena Hartulari, "Istoria vieții mele de la anul 1801," *Convorbiri Literare* 10 (1926), 845; see also
 Constanța Vintilă, *Changing Subjects, Moving Objects. Status, Mobility, and Social Transformation in
 Southeastern Europe, 1700–1850*, trans. James Christian Brown (Paderborn, 2022), 241–269.
42 Potra, "Statele Europei," 208.
43 Vasile Docea, *Relații româno-germane timpurii: împliniri și eșecuri în prima jumătate a secolului XIX*
 (Cluj-Napoca, 2000), 65.
44 Smyth, *A Year*, 29–30.

meeting with two well-mannered and witty Moldavians on board the river steamboat leaving Galați. One of them, a certain George R., was a rich landowner on his way "to Paris to see his son, who was studying there, and to care for his sick wife." The boyar had a sincere and open character, spoke French well, and hated the Russians. The other, a young merchant from Galați, was extremely elegant and refined. There were numerous other Moldavian boyars on the steamer, who spent their time in the saloon playing games of chance and drinking champagne.[45] The Czech officer Emanuel Friedberg-Mírohorský (1829–1908) conversed much on the Austrian steamboats on which he traveled in March–April 1856 between Budapest and Brăila with the Moldavian boyaress Dunca and Constanța, her "enchanting" daughter. The ladies were returning to their homeland after a stay in Paris and Vienna and made a good impression among the Austrian officers on board.[46] The Dane Frederik Schiern (1816–1882), professor of history at the University of Copenhagen, who traveled on the Danube in 1857, met on the steamboat a large group of Romanians of various ages returning from the West, either from Paris or from the German spas. Princess Natalia Ghica (1835–1899), daughter of Grigore Alexandru Ghica (1803 or 1807–1857), ruling prince of Moldavia from 1849 to 1856, was the most distinguished figure on board.[47]

In the same year, the German Richard Kunisch recalled the moment when minstrels on the steamboat began to play a *hora*. Immediately, "two Wallachian passengers on the deck jumped to their feet, took one another's hands, and began that circle dance, while the others watched them. But not for long; soon a third joined the circle, followed by a fourth, by a fifth, and each time the circle of men opened to receive the next one, without the dancers stopping even for a moment."[48]

The steamboat was a space of intense socializing, and the Romanians took maximum advantage of these few days on the river to update themselves on the latest political or society news. The voyage thus became more

45 de Perthes, *Voyage*, 417–419.
46 Emanuel Salomon Friedberg-Mírohorský, *De la Praga la Focșani. Pe Dunăre spre România. Amintiri din sejurul militar în Principatul Valah din anul 1856*, trans. and ed. Anca Irina Ionescu (București, 2015), 41, 50, 76.
47 R. V. Bossy, "Un drumeț danez în Principate," *Analele Academiei Române, Memoriile Secțiunii Istorice* III, no. 24 (1941–1942), 1–8.
48 Kunisch, *Bukarest*, 218–219.

pleasant and time passed faster. Ion Codru-Drăgușanu (1818–1884), better known by his literary pseudonym "the Transylvanian wanderer" (*peregrinul transilvan*), confessed that he had repeatedly failed to admire the Danube landscape. On one occasion, the mist had been to blame, but on an another, his attention had been confiscated by "a little she-devil of a brunette" who had come on board at Budapest. "She so fascinated me, the cursed girl, that I couldn't take my eyes off her." Nor did the prince in whose suite Codru-Drăgușanu was traveling have much leisure to admire the landscape, finding himself obliged to entertain an "elderly matron [...] of high rank."[49]

Conversation was an integral part of the voyage. In 1846, when Vasile Alecsandri traveled to Istanbul on board the steamer *Baron Eihoff*, discussions with acquaintances made the journey more pleasant: "Entering the first class saloon, I come upon Princess Mavrogheni and her daughter, with whom I spend two hours in a very pleasant conversation, swearing that we shall leap to one another's assistance in the event of shipwreck on the Black Sea."[50] On her way to Vienna in 1852, Zoe Golescu (1792–1879) conversed at length with Alecsandri and two Moldavian ladies.[51] That the discussions might be intense is clear from the remark of Constantin Hurmuzaki (1811–1869), who noted in 1860 that he wondered that the Ministry of Justice had been given to Gheorghe Crețeanu (1829–1887), as "I met him on the steamboat and found him very stupid."[52]

THE STEAMER IN ROMANIAN LITERATURE

A number of Romanian authors of the time left substantial accounts of their experiences on board the Austrian steamers: Vasile Alecsandri, Dimitrie Bolintineanu, Dimitrie Ralet, Alexandru Pelimon, and Nicolae Filimon.

---

49 I. Codru-Drăgușanu, *Călătoriile unui romîn ardelean în țară și în străinătate (1835–44)* ("peregrinul transilvan"), ed. Constantin Onciu (Vălenii-de-Munte, 1910), 73–77.

50 Constantin D. Papastate, *Vasile Alecsandri și Elena Negri: cu un Jurnal inedit al poetului* (București, 1947), 169.

51 *Din vremea renașterii naționale a Țării Românești: Boierii Golești*, vol. 3, *1850–1852*, ed. George Fotino (București, 1939), 377–379.

52 *Din relațiile și corespondența poetului Gheorghe Sion cu contemporanii săi*, ed. Ștefan Meteș (Cluj, 1939), 52.

Possessed by the "tourist demon," Alecsandri fell in love with the steam-boat and on the steamboat about which he frequently wrote. In Vienna in 1839, the poet was put on board the steamer against his will by the banker Scaramanga and sent home.[53] Later, Alecsandri sailed all over the Mediterranean on board various steamboats, and his romance with Elena Negri (c.1822–1847) was largely played out at sea.[54] The sea was also the setting for Elena's untimely end. On April 25, 1847, the two left for Istanbul on the French steamboat *Mentor*, in the company of Costache Negri (1812–1876), Ion Bălăceanu (1828–1914), and Kogălniceanu. On May 4, Elena died on board the vessel as it entered the Golden Horn.[55]

In one of his numerous voyages on the Danube, Alecsandri was witness to one of the most serious shipping accidents recorded in the period: the collision of two steamers. Given the considerable public interest in the event, he decided to write about the accident. He wrote a separate text on the subject and also included an account of it in one of his comedies, *Un salon din Iași* ("A Salon in Iași"). His talent for storytelling makes the episode extremely vivid, and it deserves to be quoted *in extenso*.

Alecsandri had embarked at Orșova on the steamboat *Széchenyi*, which sailed on the Wallachian side as far as Galați. Contrary to regulations, but relying on the helmsman's skill, the captain had decided to continue sailing even after nightfall. It was November 9, 1851, late autumn, and

the cold and damp of the night had driven the travelers from the deck, so that both the 1st and 2nd class saloon and the cabins were packed with all sorts of peoples: Romanians, Italians, English, French, Turks, Serbs, Greeks, Jews, even an Indian prince, the nabob Ecbalod-Daula-Dod, with whom I had once left Paris.

The program was the usual one: "Some of them played cards, others read, others wrote, and most talked in their various languages, thus turning the steamboat into a veritable Tower of Babel." Around midnight, "Sleep

53 Păltănea, *Viața lui Costache Negri* (Iași, 1985), 44.
54 Ioana Manta-Cosma, "O cafea, o lulea și o iubită. O iubire 'pașoptistă' în Veneția anului 1846," *Caiete de Antropologie Istorică* 19 (2011), 15–29.
55 Papastate, *Vasile Alecsandri*.

began to soften voices, to make eyes smaller, and to produce a communal yawn, which went the rounds of the saloon several times." They were to have little rest, however:

> Suddenly, around two hours after midnight, the travelers woke in fear, to a terrible roar and a terrible shaking of the steamboat! The benches, the table, the chairs were overturned, together with those sleeping on them, and the stove collapsed as if from an earthquake. The darkness was dreadful! Dreadful too was everyone's terror, for in that confusion, at once there rose up some twenty frightened cries saying in various languages: Suntem pierduți! Nous somme perdus! Siamo perduti! Kirie eleison! Jesus Maria und Iosef! Allah! Allah! Aman! Aman! Aivei! ghevalt! Vei! Vei! etc. etc. etc. Beyond our saloon, across the passage, the ladies' cabin resounded with sharp screams, with hysterics, with weeping, with spine-curdling lamentations!

In the midst of the uproar, the passengers rushed to get onto the deck, where they were better able to appreciate what was going on:

> Our steamboat was stopped in the middle of the Danube, like a wounded beast, and through the darkness we could see further away a black monster fleeing upstream. The helmsman told us it was another steamboat, which, after it had fallen blindly upon ours and smashed it, was now going away from us without giving us the slightest help. Just then, a sailor, coming from the prow, went quickly past us, saying to the helmsman: *Our people are dying!* ... and, jumping into a small boat, he made for the fore end of the steamboat.

The narrator went to help some of the injured and met passengers coming from another saloon, which they said was flooded.

> For all that, I went down into it, and the picture that appeared before me made me forget that I was stepping through water. Three men armed with axes were trying to demolish the end wall, which separated the saloon from the sailors' cabin, to save two comrades of theirs, caught and crushed among the iron wreckage of that cabin by the collision of the

steamboats. In that terrible crash the whole fore part of our ship had been smashed, and two of the crewmen were now dying crushed and broken in their beds! [...] Their faint groans could be heard through the wall, together with the bubbling of the water that was invading the cabin; and the men on this side, in the saloon, kept striking with their axes. A vain attempt! For the walls were clothed with thick plates of iron. Soon the voices of those unfortunates could no longer be heard, but only the dull noise of the waves. The Danube had drowned them both! [56]

Meanwhile, the vessel was driven by the current toward the Wallachian bank and the passengers began to be evacuated: "It was now three hours after midnight. The steamboat was deserted and the steam from the engine, coming powerfully out of the chimney pipe, produced a sad and terrible noise. It seemed like a huge beast that given up the ghost in the waves of the Danube!" The passengers were now safely landed on the lonely shore, from where

by the flame of a brightly burning fire, beside the water, there could be seen on one side the steamboat half sunk and tilted to one side, while on the other side the light fell on the different groups of travelers and the various things saved from the drowned vessel: rope, canvas, ironwork, mattresses, pillows, traveling bags, trunks, damaged plates, broken bottles, hens and turkeys tied by the legs, etc., all flung one on top of the other. Higher up, on a bed of reeds, some six ladies lay in their capes shivering with the cold; next to them, a few of the passengers were lamenting together the loss of their goods; others, further along, were walking around smoking and making jokes; others were trying to comfort one of the sailors who was weeping for his brother drowned in the cabin, and to help the unfortunate who had been injured on the head and the legs when the steamboats collided. Most, however, were crowded round the fire, boasting that that they had not been frightened at all! [57]

At Orşova, Alecsandri parted with his friend Bolintineanu, who was still banished from Wallachia, and thus could not board the steamboat from the

56 Alecsandri, "Înecarea vaporului," 73–85.
57 Ibid.

Wallachian bank. Bolintineanu embarked on the Serbian side, heading for Ruse, where he hoped to be able to see his sister. Between the geographical and historical descriptions already mentioned, Bolintineanu also found time to turn his eyes "upon the passengers with whom I found myself in this floating birdcage." Most were

> Bulgarians and Serbs, together with Turks from Bosnia. There were also some Europeans. An English lady with a woman who served her, twice the age of majority [...] There was also a young English tourist [and] a German doctor [...]. There was also a Romanian from the Banat with his wife, quite a genteel lady, [and] a French man of letters completed our society.[58]

While the travelers waited for the steamboat to arrive at Kladovo, the author made friends with the German doctor, the English tourist, and the French man of letters.[59] Bolintineanu did not return to his homeland until the autumn of 1857, when, as mentioned in Chapter 2, he traveled back from Istanbul by steamer.[60]

The Moldavian boyar Dimitrie Ralet (1817–1858) sailed by steamer from Varna to Istanbul and from there to Galați in 1855–1856. He too recalled the motley company on board the vessel: an Englishman and a Frenchman; a German violinist; Turkish women sitting "in a nest of mattresses"; a pasha "with galoshes with spurs" who dressed in *anteri* and *cüppe* but ate with a fork; "an elegant lady who had much to command, who had a maidservant, pretensions, and a lorgnette"; "ragged Jews"; and "some Bulgarians and monks who were going on *hagialâc* [pilgrimage to the Holy Places]."[61] Ralet's return journey similarly brought forth acid comments on the crowding on board, with passengers seeking "a more convenient position among the goods, Turkish mattresses, Jewish cushions and packs, English parcels, which together covered almost the whole deck." Among the first-class passengers were "a Turkish doctor, a Greek merchant, and a poor lady from Transylvania," while in the lower classes, "some Greek and Armenian petty

---

58 Bolintineanu, *Călătorii pe Dunăre și în Bulgaria*, 6.
59 Ibid., 14.
60 Bolintineanu, *Călătorii în Moldova*, 260.
61 Dimitrie Ralet, *Suvenire și impresii de călătorie în România, Bulgaria, Constantinopole*, ed. Mircea Anghelescu (București, 1979), 51–54.

merchants," and "some Jews who had been to Crimea," all playing cards "with a hum, with unrestrained vulgarity, with a tone of terrible quarrelling."[62]

Alexandru Pelimon traveled by steamer between Turnu Severin and Giurgiu in 1858. On board

> a numerous society of foreigners and natives, men and ladies, were amusing themselves by reading, eating, joking, laughing in the passengers' saloon and sometimes, as happens not uncommonly on the steamboat, disputing or quarrelling. On the cover outside, some three Muslim families sat hunched up among the mattresses, and especially the *hanıms* [ladies], sheltered under one of the vessel's tents.

Strange bonds were formed on board the steamer, "this little planet, carrying on board, in the midst of the watery element, its own world of people and goods, which it would continually change from port to port." When the weather permitted, the passengers would come out on deck

> to admire the beautiful and extensive view to be seen from afar on both shores of the Danube: the pools shining like mirrors poured on the edge of our land, those islands covered with forests, the fields laid out with crops approaching harvest time, the villages and towns along their edges, the raised banks to the right, the grand panorama of the Balkans that began little by little to be seen, a few towns suddenly appearing and the boats pulled up along the shore: an unbounded picture, an exceptional spectacle presented itself to us.[63]

Nicolae Filimon set out on his first journey abroad on June 29, 1858. After taking the omnibus from Bucharest to Giurgiu, he embarked there on the *Archduke Albert*: "The signal for departure was given, and the steamboat began first to groan like a wild bull, and then to cut the waves rapidly, although the wind was against it." Filimon also oscillated between various narrative levels and put an emphasis on his traveling companions. The private cabins and the first-class places were occupied "part of them, by a few

---

62 Ralet, *Suvenire și impresii de călătorie*, 178–179.
63 Pelimon, *Impresiuni*, 138–142.

Russian families who, wishing to take advantage of the clemency and toler-
ance of the emperor Alexander II, were going some to France and the others
to the mineral waters in Germany; and the rest were occupied by Romanian
families from both Principalities, who were also traveling for entertainment." Worthy of particular note was a provincial petty boyar couple from
Moldavia, traveling with their two daughters: "This couple were traveling
in Europe for entertainment. Their extravagances and ridiculous manners,
helped by clothing from the time of Cantemir-vodă, aroused in all the travelers a nervous laughter, made me believe that the subject of the farce entitled *Madame Chirița* is no invention but a true picture." As for the second-
class cabin, that was full of merchants going on business to *Lipsca* (Leipzig),
Paris, or London.[64]

The writers used various types of discourse, combining an informative
style, similar to that of a tourist guidebook, with a more discursive-digressive
style as they laced their accounts of their travels with anecdotes, curiosities,
and little amusing stories—and at times wrote primarily for literary effect.
As one literary critic remarked about Filimon, and the observation is equally
valid for Bolintineanu, they hesitated between the style of a Baedeker guidebook and that of a memoir, between utility (providing information), entertainment (agreeable passage of time), and the purely aesthetic.[65]

Traveling Expenses

The grand *vistiernic* (treasurer) Nicolae (Neculache) Rosetti-Roznovanu
(1794–1858) was among the leading boyars of Moldavia whose journeys
through Europe we can reconstruct. Already in 1818, he reached Vienna,
Paris, and London, and he went on to visit St. Petersburg (1825–1826), Odesa
(1828–1830), Vienna (1832–1834), Istanbul, and Chernivtsi (1842). In the
autumn of 1852, he traveled to France to enroll his son Nicolae (Nunuță)
Rosetti-Roznovanu (1842–1891) in a Paris school and also to consult with
doctors there about the father's problems. On the basis of two lists of
expenses, drawn up by the family stewards from various tickets, receipts, and

---

64 Filimon, *Escursiuni*, 14–18. Cantemir-vodă: Dimitrie Cantemir (1673–1723), scholar and prince of
    Moldavia; for Madame Chirița (*Coana Chirița*), the main character in a series of comedies by Vasile
    Alecsandri, see also Chapter 1.
65 Paul Cornea, *Aproapele și departele* (București, 1990), 163–189.

provisional lists, the historian Dan Dumitru Iacob has reconstructed Rosetti-Roznovanu's route and the payments he made.[66] We are thus able to better understand the logistics of such a journey and some of the advantages of the Danube route.

Neculache set out with a suite made up of relatives and servants. Among the family members were Nunuță, a certain *"cucoana* [Madame] Ruxăndrița" (probably the boyar's daughter-in-law), and her children, Gheorghe and Adela. Nunuță was accompanied by his tutor, Placide Doury. The party also included Rosetti-Roznovanu's domestic servants Alecu, Costache, Matei, Mihai, and Gheorghe, and Ruxăndrița was probably also accompanied by some servants. The group left Iași for Galați, via Tecuci, in carriages and Jewish carts, drawn by a number of pairs of post horses, for which the travelers paid 17 *galbeni*.[67] At Galați, they stopped at a hotel and then embarked on a river steamer, the boyars in first class and their domestic staff in second class. The tickets cost 200 *galbeni*, plus another 14 for the servants' places and food. Other expenses were incurred at Orșova and Budapest, where the *galbeni* were exchanged for Austrian florins. It is not clear whether they went from Budapest to Vienna by steamboat or by train. On September 24, 1853, they were in Vienna, where they spent a few days, before continuing on their way to Paris via Prague, Dresden, Leipzig, Frankfurt, and Strasbourg. They arrived in the French capital at the beginning of October. Their return journey to Iași was through Galicia and Bukovina, first by railway and then by the diligences of the imperial postal service or its separate coaches as far as Chernivtsi. From there, they returned to Iași in carriages and Jewish carts.[68]

We also have detailed information about a journey in the summer of 1857 to Istanbul, where Neculache met his wife Maria (Marghiolița) and one of their daughters, probably the youngest, Emma. The boyar family was accompanied by "Miss Parlet" the governess, "Fransua," and Ichim. They stayed in hotels in Galata and Pera and visited some of the tourist attractions of the city. The list of expenses shows that they saw Hagia Sophia, went to the theater

---

66 Iacob, "Călătoria lui Nicolae Rosetti-Roznovanu la Paris, în 1853," *Analele Științifice ale Universității "Alexandru Ioan Cuza" din Iași, s.n. Istorie* 63 (2017), 349–397.

67 *Galbeni* is the name for the foreign (mostly Austrian) golden coins in use in the Principalities. Rates varied a lot, but on average an Austrian *galben* (ducat) was worth, in the 1850s, 37 lei.

68 Iacob, "Călătoria."

and to casinos, and walked by the "Sweet Waters of Asia." They spent a considerable time shopping: copperware for the kitchen, Persian and Samac carpets, silk, felt, oilcloth, glass for windows of various sizes. However, it seems that the principal purpose of their stay there was medical, as Maria Rosetti-Roznovanu took a balneary treatment.[69]

According to the accounting records, the traveling expenses were considerable. For Neculache's journey to Istanbul he paid 4 *galbeni* for "the coach in which I went to Galați," 2 *galbeni* to obtain a passport, 78 lei to send a telegraph letting his wife know that he was leaving Moldavia, 86 lei for bed and board in an inn at Galați, 485 lei for the steamer ticket, 63 lei for food on the steamer, and 6 lei for "the cart for my trunk on the steamer." The total expenses for the journey between Galați and Istanbul were 1,004 lei, which pales into insignificance beside the 70,211 lei spent during the stay in the Ottoman capital. These included the cost of his return to Moldavia: 4,288 lei for seven steamboat tickets (three in first class and four in second class) and 1,135 lei for "sundries on the steamboat." There was also 506 lei for "food on the steamboat 7 persons on the return," 69 lei for "tips to two maids," 20 lei for "the cart for the lady's things, to the hotel," and other expenses incurred at Galați for the acquisition of rice, sugar, "chests, one of wood from Provence," candles, marinated sturgeons, marinated crayfish, sardines, and citrus fruits.[70]

The previous year, 1856, Neculache and Marghiolița had traveled to Vienna and Paris, probably for reasons concerning, at the same time, their son's education and their own health. In the autumn of 1857, seriously ill with dropsy, Nicolae Rosetti-Roznovanu embarked once more at Galați, with a reserved cabin on the river steamboat. George Sion, going abroad on his honeymoon, noted that he had conversed all the way with Lady Marghiolița, whom he had not seen for a long time: "Then, as if she needed an intimacy, she started first to ask me about my life and then told me about hers. [...] All the way as we traveled to Pest, she never ceased to greet me and to speak to me with much affection."[71] The boyar Rosetti-Roznovanu's illness worsened, and he died in Vienna in May 1858.

---

69   Corina Cimpoeșu, "Călătoria la Constantinopol în anul 1857 a familiei Rosetti-Roznovanu," *Ion Neculce* 13–14 (2007–2009), 37–45.

70   Ibid.

71   Gh. Sion, *Suvenire contimporane* (București, 1915), 388.

The Revolutionary Steamboat

The Bulgarians' movement for national liberation is closely linked to the steamboat *Radetsky*, built in 1851. It was operating on the Lower Danube routes in 1876, when, during the Balkan conflict, it was hijacked by the Bulgarian revolutionary group led by the poet Hristo Botev (1847–1876). On May 29, 1876, after the steamer had left Bechet, the insurgents, who had boarded under cover at various Danube ports, forced the captain, Dagobert Engländer, to take them to Kolzloduy, where they attempted to launch the revolution of national liberation. The *Radetsky* remains in the collective memory of the Bulgarians as an instrument of revolutionary struggle, aided by the fact that a replica of the steamboat now functions as a museum.

Less well known is the fact that the Austrian steamers also played an important role in the progress of the revolution in the Romanian lands. Among those who took full advantage of the mobility they offered on the Danube were young men with progressive ideas. As neutral vessels operating on an international waterway, the steamers were used by the revolutionaries to return to their homeland or, if necessary, to escape from it.

Thus, it was in 1848 that a large part of the Paris-based revolutionary nucleus arrived in the Principalities via Vienna, where they embarked on the steamer. An interesting meeting between those arriving and those escaping is vividly recounted in the memoirs of Radu Rosetti (1853–1926):

> When [the future prince Grigore Alexandru Ghica] was returning from Vienna, where he had gone to bring my mother from the pensionnat back to this country, the steamboat on which they were passed, at Giurgiu, that on which my grandfather Lascăr [Rosetti] and the other exiled Moldavians, who had escaped Miscenko's escort at Brăila, were going to Baziaș. Grigore Ghyka, who had found out on the way about what had happened at Iași, called out to the steamboat coming from downstream, which had pulled in close to that on which he was: "Isn't there on this steamboat any wandering Moldavian?" And they, his friends and comrades in the struggle, answered at once, and thus they were able to exchange a few words from steamboat to steamboat.[72]

---

72   Radu Rosetti, *Amintiri. Ce am auzit de la alții. Din copilărie. Din prima tinerețe* (București, 2017), 220.

From other sources we learn that in April 1848 Costache Negri, Iancu Alecsandri (1826–1884), Dimitrie Bolintineanu, and Alecu Russo (1819–1859) left Vienna for the Principalities. They were not allowed to enter their homeland; so they reembarked on the steamboat going upstream.[73] There they found a number of Moldavian revolutionaries who had been arrested in Iași the previous month and had been held in Galați with the aim of banishing them to the Ottoman Empire. A plot hatched by the young men's families, with the support of the British and Austrian vice-consuls, had resulted in some of the arrestees being rescued at Brăila—among them Alexandru Ioan Cuza (1820–1873), Manolache Costache Epureanu (1823–1880), Alecu Moruzi (1815–1878), Lascăr Rosetti (1816–1884), and Zaharia Moldoveanu (1820–?). Safely embarked on the steamboat *Franz I*, Cuza wrote to consul Huber to thank him, in the name of his comrades, "for all the care that the captain of this steamboat" had shown them, at the consul's recommendation. At Giurgiu, the group of fugitives met their fellow revolutionaries who had been refused permission to enter the country, and together they continued their journey upstream. On reaching the Banat, some of them headed for Lugoj, while Cuza and Cantacuzino left for Budapest, where they received an enthusiastic welcome, and all continued from exile their campaign against the regime of the ruling prince of Moldavia, Mihail Sturdza. The Wallachian revolutionary Nicolae Bălcescu (1819–1852) also embarked on a steamboat in Vienna and then continued overland from Orșova to Bucharest.[74] Indeed, the Catholic priest Jacques Mislin, a theologian with conservative convictions, who was traveling on the Danube at the end of June 1848, noted that on the steamboat he had met young boyars who sang the Marseillaise and spoke of "universal enfranchisement, progress, and socialism."[75]

Once in power, the revolutionary government in Bucharest used the steamboat service to promote its cause through diplomacy, but also to hamper the mobility of its political adversaries. In August 1848, the government tried to prevent the return of Colonel Ioan Odobescu (1793–1857), who intended to embark on the steamboat *Árpád*, which circulated on the Wallachian side. The prefect of Mehedinți found out about the plan and invited "the Agent of the steamboats in future not to permit to embark in

73 Păltănea, *Viața*, 91–92.
74 Sion, *Suvenire*, 228.
75 Mislin, *Les Saints Lieux*, 50.

the steamboats any passenger, before seeing on their passports the visa of the Administration." Odobescu protested, considering that the Wallachian authorities had no right "to give orders on the steamboat, since it is floating on water, and not on dry land."[76]

The episode is well known in which the arrested revolutionaries in Wallachia were shipped to Ruse and from there to Vidin. Eventually, after numerous adventures, they left by steamboat for Zemun. As Bălcescu wrote to his friend Ion Ghica, "Others may have thrown themselves into the Danube, [but] I threw myself onto the Danube to go to Vienna."[77]

The crews of the steamboats also included Romanians with revolutionary aspirations, according to Ion Ionescu de la Brad. Such was the case of a certain Rălescu from Mehadia; "the pilot too is Romanian, Dumitrașcu, from Wallachia, a sturdy man and with the greatest national enthusiasm. They struck with all their hearts against the servitude in which the peasants have fallen, with the doubling of the tithe and of the working of the two acres."[78]

Banished to the Ottoman Empire or elsewhere, the revolutionaries could remain close to their homeland through the intermediary of the steamboat. We may recall Bolintineanu's journey in 1851, and similarly Bălcescu tried to see his family again before his premature death.[79] The *spătar* Sandu (Săndulache) Miclescu (1804–1877) was living in Izmir at the beginning of 1849 and later in Istanbul. Through letters sent to his homeland, he managed to deal with matters concerning the management of his estate. At the same time, he invited his wife to join him: "But if you can and want to come here, I await you with impatience. On the steamboat you will take the first class for yourself—the child being small enters without payment—and your maidservant on the deck, taking food for her from Galați, so as not to pay too dearly on the steamboat."[80] Zoe Golescu also set out for Istanbul in the hope of seeing her exiled sons and obtaining their freedom.[81]

---

76 *Anul 1848 în Principatele Române: acte și documente*, vol. 3 (Bucuresci, 1902), 347–348, 563, 599.

77 Nicolae Bălcescu, *Scrisori către Ion Ghica*, ed. Petre V. Haneș (București, 1911), 120.

78 Slăvescu (ed.), *Corespondența*, 97.

79 Cornelia C. Bodea, "Călătoria lui Bălcescu pe Dunăre în 1852," *Studii* 10, no. 1 (1957), 161–170.

80 Mihai-Cristian Amăriuței and Benone Dorneanu, "Spătarul Sandu Miclescu de la Șerbești (1804-1877). Schiță de portret a unui 'pașoptist' uitat, pe baza unor documente și amintiri de familie," *Archiva Moldaviae* 10 (2018), 82.

81 Elisabeta Gheorghe, "Între etic și estetic. De ce călătoresc româncele? De ce scriu," *Studii și cercetări științifice. Seria filologie* 34 (2015), 44.

This opening to the world also had its opponents, as can be seen from the reproach of an elderly boyar, in whose view the steamboats brought to the Principalities not only "civilized" persons but also vagabonds or elements ready to corrupt the nation: "Don't you see that each steamboat comes to Galați laden with regiments of the dissolute and the crazed; you young people will see what consequences all this will have for the future of the poor country."[82]

Conclusions

From the 1830s to the 1860s, the Austrian steamboats provided the Romanians with their most important connection to the rest of the world. Continuing a period of opening in which the number of travelers to the East and the West alike had been continuously growing, the introduction of the steamer service between Vienna and Istanbul placed the Principalities on a popular international route. Linking Wallachia and Moldavia to two great European capitals, transport hubs offering access to a variety of other destinations in the West and in the Orient, the steamers made a profound contribution to the economic, social, and cultural modernization of the Romanian space. On them, not only travelers but also goods, information, and ideas circulated, all of which were vital for the way in which the Romanians adapted to the game of modernity.

The steamboat became part of normality for those Romanians who, for pleasure or out of necessity, went to get to know the wider world. For these mobile representatives of society, the steamer cabin was a fascinating space of socialization in the manners of "civilized Europe." The saloon rapidly incorporated the Romanians in the company of global tourists, allowing them to mix with the high (and not-so-high) society of Europe, where they could now arrive more rapidly, more cheaply, more safely, and more comfortably.

---

82 Costachi Konachi, *Poesii: alcătuiri și tălmăciri*, 2nd ed. (Iași, 1887), 89–90.

CHAPTER 5

Travels and Epidemics[1]

A Wallachian Tragedy

On the evening of August 12, 1849, tragedy struck the family of Barbu Știrbei (1799–1856), ruling prince of Wallachia from 1849 to 1856. The newly enthroned prince was returning home after his investiture in Istanbul, and a number of Wallachian notables were preparing to welcome him in the port of Brăila. Protocol decreed that when the steamer arrived Știrbei "would get off under the shade made for the purpose, where he would receive all the authorities. After that he would go to the quarantine premises, which had been emptied of their employees and prepared for the accommodation of the Prince and his suite."[2] Also present in the Wallachian port was the inspector general of quarantine for the two Principalities, General Nicolae Mavros (1782–1868), who gave orders that the sanitary regulations were to be strictly followed. As the Austrian steamer had arrived in the port after sunset, the procedures required that the passengers should not disembark until the following morning. The prince's adjutants, among them his son-in-law, Major Alexandru Villara (?–1849), did not react well to the restriction, being "bored by the sea journey and annoyed that in their own country they could not get out." Accounts of what followed vary, but what matters most for our purposes is the denouement: Villara fell into the Danube and, despite desperate attempts to rescue him, he was found drowned. There followed, the

---

1 Additional research for this chapter was also supported by a grant of the Romanian Ministry of Education and Research, CNCS – UEFISCDI, project number PN-III-P4-PCE-2021-1374, within PNCDI III.
2 Grigore Lăcusteanu, *Amintirile colonelului Lăcusteanu*, ed. Rodica Pandele Peligrad (Iași, 2015), 207–212.

next day, the prince's stern rebukes directed at the inspector general, who, in his turn, blamed the quarantine director, Schina, who should have reported to Știrbei and invited him to the "quarantine houses" as soon as the steamer docked. "After three days of quarantine completed," continues the eyewitness Grigore Lăcusteanu (1813–1883), a source far from favorable to Mavros, the prince continued on his way to Bucharest.[3]

Over and above any element of pride and political frustration, General Mavros's excess of zeal with regard to the princely entourage served to underline the seriousness and strictness of the quarantine provisions, in the face of which all travelers were to be treated with equal attention. The disease spared no one and, especially in times of epidemic, the severe precautions were applied to *all* those who arrived from possible hotspots of infection. The lazarettos were in the frontline of the struggle against disease, and scrupulous respect for the sanitary procedures was vital for the security of the state. Travelers might be discontented at the rigidity of the regulations, but public health was more important than any temporary personal discomfort. Villara's death took place in regrettable circumstances, but respect for the epidemiological provisions meant that no one could be considered to blame for the unfortunate accident. In an attempt to explain the operations of the quarantine regime set up along the Lower Danube, I shall examine in this chapter the "confrontation" between the popular route of travel on the Danube and the system of quarantine imposed by the states on the great river's banks. Among the Danube's many roles, especially in the decades before the Crimean War, was that of a symbolic frontier between different epidemiological regimes or, as historian Bogdan Popa notes, cultural and mental spaces in which health and sickness were perceived differently.[4]

With outbreaks of disease becoming more and more numerous and more virulent in the region, against the background of the rapid growth in the circulation of people and goods along the Danube, a traveler between the Iron Gates and Sulina would come in contact with four approaches to the nature and transmissibility of epidemics, to be seen in

---

3   Ibid. Details in *Mărturii istorice privitoare la viața și domnia lui Știrbei Vodă*, ed. Nicolae Iorga (București, 1905), 8–9.

4   Bogdan Popa, "Experiența fizică a frontierei: carantina," in Romanița Constantinescu (ed.), *Identitate de frontieră în Europa lărgită: Perspective comparate* (Iași, 2008), 93.

the organization and function of the quarantine services on the territories of the Habsburg Empire, the Ottoman Empire, the Principalities of Wallachia and Moldavia, and imperial Russia. In the Danube lazarettos, which I shall examine as transimperial "contact zones," travelers became familiarized not only with the severe procedures to combat disease but also with policies of controlling the mobility of ideas that threatened the political health of the respective states.

The experience of quarantine isolation was an integral part of travel in the nineteenth century. From the 1830s to the 1850s, a time that saw accelerated development in international tourism, but also the global spread of devastating epidemics, lazarettos and periods of quarantine "arrest" figure prominently in travel narratives. The consumption of such writings was also considerable as the public was interested in diseases and how they spread, in medical institutions and practices, and in forms of epidemiological isolation and exclusion. The American philologist Kelly L. Bezio has even identified the "quarantine narrative" as a literary subgenre.[5] In free-standing texts or as part of more extensive accounts, writers recorded the details of their experience, referring to infection and purification, captivity and closed spaces or the global space constituted by a lazaretto. In what follows, I shall describe the organization of quarantine on the Danube and the experiences of a number of travelers in the space and time of the lazaretto, on the basis of such pages in the literature of sanitary detention.

Historians have investigated the issue from various perspectives. In this chapter, I shall make use both of the findings of the "classic" historiography of quarantine, according to the approaches to medical and political history of Gunther E. Rothenberg,[6] Daniel Panzac,[7] and Erwin Ackercknecht,[8] and of newer studies of a Foucauldian character and of transnational history, with reinterpretations that bring under discussion the biopolitical technologies used in the lazarettos and the role of these institutions in projecting the

---

5   Kelly Bezio, "The Nineteenth-Century Quarantine Narrative," *Literature and Medicine* 31, no. 1 (2013), 63–90.

6   Gunther E. Rothenberg, "The Austrian Sanitary Cordon and the Control of the Bubonic Plague: 1710–1871," *Journal of the History of Medicine and Allied Sciences* 28, no. 1 (1973), 15–23.

7   Daniel Panzac, *La Peste dans l'Empire ottoman 1700–1850* (Leuven, 1985).

8   Erwin H. Ackerknecht, "Anticontagionism between 1821 and 1867: The Fielding H. Garrison Lecture," *International Journal of Epidemiology* 38, no. 1 (2009), 7–21.

power of the state and in the construction of the national space.[9] Analyzed
as an instrument of control of borders and management of the migration of
people and goods, as an institution where sanitary and fiscal documents were
issued and checked, or as a service for the collection of information about
wrongdoers or political enemies, the quarantine system constituted a classic
type of "dispositif" for the manifestation of the power of the state.[10]

THE CORDON SANITAIRE IN THE HABSBURG TERRITORIES

The frontier between the Habsburg and the Ottoman territories was defended
by border regiments whose role was as much sanitary as it was military. Well
organized already in the eighteenth century, the regiments of border guards
ensured the protection of the cordon sanitaire, the defensive mechanism
intended to reduce the circulation of epidemics coming from the potentially contaminated space of the Orient. As a doctor in the Austrian service
remarked in 1847, the quarantine stations were "floodgates and guard posts on
the Turkish border, to protect Europe from plague."[11] The system of prevention was based on the model already applied in earlier centuries in the ports of
the Mediterranean, now adapted to function along a terrestrial border.

The quarantine stations or lazarettos were the points that filtered mobility
from Ottoman to Habsburg territory. All that entered from the Balkans—
travelers, animals, vehicles, and goods—had to be subjected to certain procedures to mitigate the risk of importing epidemics. The principal factor in
anti-epidemiological prevention was time: the length of time travelers spent
in quarantine had to be sufficient to allow symptoms of the illness to appear
before they passed through the cordon sanitaire. Other factors contributed
to a gradual reduction from the period of waiting originally recommended,
namely forty days (*quarantena*). Good management of information, with

---

9   Peter Baldwin, *Contagion and the State in Europe, 1830–1930* (Cambridge, 1999); Mark Harrison, "Dis-
    ease, Diplomacy and International Commerce: The Origins of International Sanitary Regulation in the
    Nineteenth Century," *Journal of Global History* 1, no. 2 (2006), 197–217; Alison Bashford (ed.), *Quar-
    antine: Local and Global Histories* (Basingstoke, 2016); John Booker, *Maritime Quarantine. The British
    Experience, c. 1650–1900* (London, 2016).
10 Andrew Robarts, *A Plague on Both Houses? Population Movements and the Spread of Disease across the
    Ottoman-Russian Black Sea Frontier, 1768–1830s*, PhD thesis (Georgetown University, 2010), 222.
11 Ion Negru, "Cum vedea doctorul Pavel Vasici carantinele în 1847," in G. Brătescu (ed.), *Din istoria luptei
    antiepidemice în România. Studii și note* (București, 1972), 315.

credible date on the sanitary situation in the territories from which travelers came, enabled the quarantine period to be reduced in periods when there were no active epidemics in the Ottoman territories. Decontamination, according to the scientific knowledge of the age, visual inspection of travelers' bodies by medical personnel, and the segregation of travelers for the duration of their quarantine detention completed the list of measures by which the authorities sought to diminish the risks of epidemic.

A major challenge for all the cordons sanitaires was how to balance blocking the spread of disease with the need to keep the border open for trade, which was seen, in a rationalist spirit, as a factor of human well-being. Along the frontier, from the Adriatic to Bukovina, sixteen quarantine stations were thus established. Relevant for this chapter is the lazaretto of Jupalnic, situated on the Danube, close to Orşova in the Banat.

In the nineteenth century, against the background of the Habsburg economic offensive against the Ottoman provinces, there was a perceptible tendency to reduce the period of quarantine detention for travelers and to simplify the procedures for cleaning goods. This resulted also from the proactive attitude of the Austrians toward the epidemiological situation in the southeastern Europe. The security of the cordon sanitaire was the result of a better management of information, in line with the efficiency of the imperial bureaucracy and with the introduction of new means of communication. The system needed reliable information fast, or else significant economic loss would result. The speed at which information circulated and the quality of that information were coming to be of vital importance, and the Austrians mobilized the Sublime Porte in this respect, persuading them to become an integral part of the transnational struggle against the spread of epidemics.

The Habsburg quarantine system played an important role in the modernization of the sanitary facilities of the Ottoman lands, including the two Principalities. Starting in the 1840s, Austrian doctors were officially involved in the reform of the medical structures of the Ottoman Empire. In taking on this "civilizing" role in relation to the medical institutions of the Levant, the authorities in Vienna were in fact seeking to improve their own sanitary protection by transferring part of the cordon sanitaire southwards. In a period when the utility of quarantine was questioned by many specialists in Austria and in the Western world, the Viennese authorities encouraged preventive

sanitary measures in the Ottoman Empire and supported this effort through
the transfer of medical knowledge and the creation of institutional structures to fight against disease. Carl Ludwig Sigmund (1810–1883), one of the
most renowned Austrian experts on contagious diseases, was of the opinion
that Europe's line of defense against epidemics, which the Habsburgs had
successfully maintained for some centuries, had to be moved further south.
In campaigning for a change in quarantine policy, Sigmund maintained that
in the long term the fight against epidemics would have to be fought in the
areas where the diseases originated—in India or Egypt. With Austrian support, a medical school was established in Istanbul, where sanitary staff were
trained who would later play a role in combating the spread of epidemics.
The Austrian cabinet repeatedly modified the period of quarantine arrest
according to the epidemiological situation in the East, so that by the end of
the 1840s the quarantine had been considerably reduced.[12] In the early 1850s,
when there were no epidemics active in the region, quarantine inspection
was a mere formality.

As mentioned above, the Austrian lazaretto that was important for
navigation on the Lower Danube was that at Jupalnic, close to Orşova,
where travelers entering the Austrian territories by way of the Danube
underwent quarantine. We shall look more closely at it in the following
sections.

On Ottoman Lazarettos

Ottoman sanitary organization in the years following the Treaty of
Adrianople was accelerated by the ravages of the cholera epidemic of the
1830s. Under pressure from the European powers to heighten their efforts
to combat the spread of the disease, the authorities in Istanbul initiated
epidemiological supervision measures, first in the capital itself and then
at other points of entry to the empire. By the end of the 1840s, there were
around eighty quarantine stations, including those in the Danube ports of
Tulcea, Măcin, Silistra, Ruse, Svishtov, Nikopol, and Vidin. By this time,
the Sublime Porte had modern sanitary regulations, aimed at preventing

---

12 Marcel Chahrour, "'A Civilizing Mission'? Austrian Medicine and the Reform of Medical Structures in
the Ottoman Empire, 1838–1850," *Studies in History and Philosophy of Science Part C: Biological and Biomedical Sciences* 38, no. 4 (2007), 687–705.

epidemics from spreading in the first place, rather than merely reacting to them.[13]

In 1839, a Superior Health Council was set up in Istanbul, comprising delegates of European states alongside its local members. The ground was thus laid for closer collaboration with a view to the standardization of anti-epidemic procedures. This process continued through the organization of international sanitary conferences, the first of which was held in Paris in 1851. Representatives of twelve states, including the Porte, took part, which favored the opening of a new front in the transnational struggle against epidemics, with the establishment of common rules of public hygiene, sanitary inspection, and quarantine organization.[14]

The interest of the European powers in the sanitary situation in the Ottoman Empire resulted both from the acceleration of commercial exchanges with the Levant and from the fact that, by virtue of its geographical position, the empire lay on the main routes along which epidemics spread. Prejudices associating Oriental societies with the presence of disease also played their part. As mentioned above with reference to the Habsburg Empire, the support of the European powers was a determining factor in the preparation of human resources and sanitary procedures to combat the spread of epidemics. This international support led some faithful Muslims to see the quarantine system as a foreign imposition. There was particular opposition to certain decontamination measures, on the grounds that they were contrary to the principles of Islam. In 1840, for example, several hundred Muslim women protested against the introduction of quarantine in the port of Varna, threatening the director of the establishment.[15]

It was in Istanbul itself that travelers most frequently encountered the Ottoman quarantine system. Along the Danube, the lazarettos on the right (Turkish or Bulgarian) bank were especially active in periods

13   Robarts, *A Plague on Both Houses?* 212–214; Aytuğ Arslan and Hasan Ali Polat, "Travel from Europe to Istanbul in the 19th Century and the Quarantine of Çanakkale," *Journal of Transport & Health* 4 (2017), 10–17.
14   Nermin Ersoy, Yuksel Gungor, and Aslihan Akpinar, "International Sanitary Conferences from the Ottoman Perspective (1851–1938)," *Hygiea Internationalis* 10, no. 1 (2011), 53–79. More details in Birsen Bulmuş, *Plague, Quarantines, and Geopolitics in the Ottoman Empire* (Edinburgh, 2012) and the larger context in Booker, *Maritime Quarantine*, 481–516.
15   Christian Promitzer, "Prevention and Stigma: The Sanitary Control of Muslim Pilgrims from the Balkans, 1830–1914," in John Chircop and Francisco Javier Martínez (eds.), *Mediterranean Quarantines, 1750–1914. Space, Identity and Power* (Manchester, 2018), 148–149.

when epidemics haunted the region. However, due to the arrangements regarding circulation along only one bank of the river, passengers on the Austrian steamboats did not undergo quarantine upon disembarkation in the Ottoman Danube ports, so there are no references to sanitary policies in their travel accounts.

Quarantine Autonomy and Antiepidemic Protectorate

The quarantine situation in Wallachia and Moldavia was also regulated in this period.[16] The Peace of Adrianople was signed precisely when a virulent plague epidemic had hit southeastern Europe. The Russian negotiators called for significant privileges to the granted to the Principalities, including quarantine autonomy. Wallachia and Moldavia were separated from the Ottoman Empire, in a form of territorial individualization that gradually extended from quarantine to other areas. The Principalities became an anti-epidemiological outpost of Russia, a buffer territory that would serve to defend the Russian Empire against its most terrible enemy—disease. Just as the Principalities' political autonomy was subordinated to a form of often inconvenient protectorate, likewise their quarantine autonomy depended on Russia, which had reserved the right to appoint a sanitary representative of the tsar's government, an official with considerable influence in the internal politics of the two autonomous states.[17]

Nicolae Mavros, whom we met at the beginning of this chapter, is one of the most interesting figures in the nineteenth-century Romanian national renaissance. Of Greek origin, he was private secretary to the last Phanariot ruler of Wallachia, Alexandru Suțu (1758–1821, last of several periods of rule: 1818–1821) and seems to have been involved in the actions of the Greek revolutionary society Filikí Etería. After the debacle of the movement in 1821, Mavros took refuge in the Russian Empire and made a career in the tsar's army, where he rose to the rank of general. He returned to the Principalities in the entourage of General Pavel Kiselyov (1788–1872), who

16 Details in Lidia Trăușan-Matu and Octavian Buda, "Cholera, Quarantines and Social Modernisation at the Danube Border of the Ottoman Empire: The Romanian Experience between 1830 and 1859," *Social History of Medicine* 36, no. 1 (2023), 24–41.

17 Robarts, *A Plague on Both Houses*, 211, states, citing documents from the Turkish archives, that there were also Ottoman officials in the Principalities' lazarettos. I have not identified any data on such a presence in the available sources.

appointed him inspector general of quarantine, first in Wallachia and then also in Moldavia. His activity was decisive for the standardization of quarantine practice in the two states—an important step, due to the medical, administrative, economic, and political aspects it involved, toward their later economic and political unification. Mavros remained in this post until the 1850s, and many of his contemporaries considered that his true loyalty was to Russia, not to the two Principalities. Today he is best known as a collector of antiquities, later donated to the National Museum in Bucharest, but his great achievement was the creation of a modern and efficient quarantine system.

The quarantine service was minutely organized under the Organic Regulations.[18] In each Principality, a sanitary committee—itself composed of a medical committee and a management committee—was set up, charged with coordinating public health policy and the maintenance of quarantine stations. All along the Danube border, from Vârciorova to the Prut, Wallachia and Moldavia were protected by a cordon sanitaire, made up of troops who formed the nucleus of a national militia, constituted, as a contemporary noted, for "the service of the quarantines, of the customs posts, and of internal order."[19] Thus, the quarantine stations fulfilled several roles, as points of sanitary control of the mobility of people and goods, but also as a "border police" and a "customs directorate." These functions are still carried out today, at any border crossing point, by various national institutions. It is important to remember that in the 1830s it was sanitary security that lay at the foundation of an extensive process of administrative modernization in the two Principalities.

All along the line of the Danube, the cordon sanitaire was patrolled by pickets entrusted with guarding the riverside communities. This service, comparable to that of the Grenzer regiments of Transylvania and the Banat, released the peasants along the cordon from army conscription and other obligations. Each picket was under the orders of a corporal and a private soldier. Wallachia had a total of 217 pickets, and Moldavia 15.[20] Adapting

18 Paul Negulescu and George Alexianu, *Regulamentele Organice ale Valahiei și Moldovei* (București, 1944), 79–85, 279–290.
19 Jean Alexandre Vaillant, *La Roumanie ou histoire, langue, littérature, orographie, statistique des peuples de la langues d'or, ardialiens, vallaques et moldaves, résumé sous le nom de Romains*, vol. 3 (Paris, 1844), 74.
20 Ibid., 46.

Daniel Panzac's observation regarding the Ottoman Empire, we may say that sanitary policy contributed substantially to the stabilizing of borders and the territorial delimitation of the two Principalities.[21] The modernization of their institutions was also largely due to fears regarding epidemics.[22]

As points of connection with the Turkish or Bulgarian bank of the river, eleven quarantine stations or lazarettos were organized in Wallachia. These were divided into three categories, according to the importance of their respective localities for exchanges with the opposite bank. The first-class quarantine stations were at Brăila, Giurgiu, and Călărași; then there were five establishments in the second class (Turnu Severin, Calafat, Turnu Măgurele, Zimnicea, and Oltenița) and three in the third class (Izvoarele, Bechet, and Gura Ialomiței). In Moldavia, there was a single quarantine station: that of Galați (first class).[23] The first-class quarantine stations were the most important for the mobility of individuals and goods on board the Austrian steamers, and it is to those that I shall refer in the following sections.

Despite the introduction of the system of Danube quarantine, the Austrians still situated the Principalities in the zone of epidemiological uncertainty. Thus, travelers coming from Moldavia and Wallachia had to undergo a period of quarantine on entering the Habsburg territories, even if they had already done so when they entered the Principalities. The measure was due to lack of trust in the efficient functioning of the quarantine system of the Principalities, until their procedures and operations were consolidated on the basis of firm medical principles.[24]

SANITARY FORTRESSES IN THE DANUBE DELTA

Under the Treaty of Adrianople, Russia had annexed the Danube Delta, and the new Russian–Ottoman frontier thus followed the course of the Maritime Danube and of its southern branch, the St. George. The treaty established the neutrality of the border region, where no fortifications could

21 Panzac, "Politique sanitaire et fixation des frontières: l'exemple Ottoman (XVIII-XIX siècles)," *Turcica* 31 (1999), 87–108.
22 Călin Cotoi, *Inventing the Social in Romania, 1848–1914: Networks and Laboratories of Knowledge* (Leiden, 2020), 93–99.
23 Georgeta Penelea, "Organizarea carantinelor în epoca regulamentară," *Studia Universitatis Babeș-Bolyai* 14 (1969), 29–41.
24 Negru, "Cum vedea doctorul Pavel Vasici carantinele," 317.

be erected: the only constructions permitted on the islands of the Delta were the quarantine stations necessary in the struggle to stop the transmission of epidemics. However, the navigable course of the river between the ports of Galaţi and Brăila and the Black Sea followed the Sulina branch, with the result that a large part of the route used by passenger and goods vessels passed through Russian territorial waters. The Russian authorities were therefore faced with the decision as to which should be guarded first: the border of the empire, along the St. George branch, or the navigable route used by foreign shipping, along the Sulina branch.[25]

In 1836, when navigation on the Maritime Danube was undergoing considerable development, including the introduction of steam vessels, Russia reorganized the quarantine service in the region of the Danube mouths. The former quarantine "border" on the northern (Kilia) branch was abolished, and the new cordon sanitaire followed the most circulated branch of the Danube, that of Sulina. There was also a lazaretto for travelers interested in passing their period of quarantine at Sulina before heading overland to Izmail. At the same time, the Russian customs service, port authorities, and river police, invoking the strict application of quarantine rules, obtained the right to inspect vessels going up the Danube to inland ports.[26]

The establishment of quarantine controls along this much-used route gave rise to numerous disputes with the captains of commercial vessels transiting the region, in general on their way to the ports of the Principalities. It was a long-term conflict, which depended also on the ambiguity of the region as a Russian territory situated along an international waterway. In a period when international law was still in its infancy, it was not entirely clear which status should prevail. However, the establishment of the quarantine service and the imposition of strict sanitary control regulations were seen as political actions aimed at discouraging the commerce of the Principalities, which were gradually freeing themselves economically from Russian tutelage. As Russophobe contemporaries saw it, under the pretext of ensuring public health, imperial Russia was erecting veritable sanitary "fortresses," by means of which it could control, in peacetime, a region of great strategic value. In the period

---

25 Details in Ardeleanu, *International Trade*, 141–148.
26 Andrei Emilciuc, "The Trade of Galaţi and Brăila in the Reports of Russian Officials from Sulina Quarantine Station," in Ardeleanu and Andreas Lyberatos (eds.), *Port Cities of the Western Black Sea Coast and the Danube: Economic and Social Development in the Long Nineteenth Century* (Corfu, 2016), 63–93.

that followed, the Russian authorities responded constructively to their critics, trying to limit the protests of Western merchants and shipowners. It should also be mentioned that, thanks to diplomatic agreements between the Habsburg and Russian governments, the Austrian steamers enjoyed special treatment. They were relatively protected from the abusive practices the captains of commercial vessels complained of.[27]

QUARANTINE: TIME, SPACE, ORGANIZATION

The period of quarantine varied according to the sanitary situation in the Ottoman Empire. In the 1830s and 1840s, it ranged from a single day to fourteen days. The tendency was toward shortening, under the pressure of growing human and commercial mobility. Depending on the information received from the Levant region, the quarantine authorities would toughen or relax the sanitary measures. Information, credible and punctual, became vital for the good running of the quarantine system and the predictability of travel times.

The time spent in the lazaretto had to be sufficient for the procedures of sanitary cleansing and political surveillance. The precautions to be taken on the arrival of passengers and goods were clearly specified in the sanitary regulations, which were similar throughout the region: passengers from the same vessel were taken to a fumigation room where they took off their "traveling" clothes and put on clean garments, generally provided by the administration of the establishment. The "detainees" were then conducted to the rooms where they were to spend their period of quarantine; their own clothes were returned to them the following day, after "purification." The travelers could not have contact with those who had arrived before or after them, so the quarantine establishment was compartmented into clearly delimited zones for each group of passengers.

Each quarantine station consisted of a series of buildings in which specific sanitary, customs, and administrative activities were carried out. These included accommodation for the resident staff, "quarantine houses" for travelers, annexes for the disinfecting of passengers and goods, stables, and so

---

27 Manfred Sauer, "Österreich und die Sulina-Frage, 1829–1854," *Mitteilungen des Österreichischen Staatsarchivs* 40 (1987), 185–236; 41 (1990), 72–137.

on. The complexity of the activities that went on in an establishment of this kind called for a diverse staff: doctors, administrative officials, interpreters, servants, officers, wardens. At Giurgiu, where one of the largest quarantine stations in the Principalities was set up in 1831, the staff consisted of a director, a subdirector, a doctor, a midwife, an interpreter (who had to speak Romanian, Turkish, and a "European" language), a secretary, two chancellery officials, four servants charged with "caring for, airing, and fumigating people and objects," together with twelve servants for the goods stores. Pay and material costs for all these amounted to 46,000 lei per annum. The quarantine employees were proposed by the Ministry of Internal Affairs and confirmed by the ruling prince. On taking up their posts, they took an oath and even needed a guarantor to vouch for their good behavior and correctness.[28] The staff of the quarantine station of Brăila in 1832 was made up of a director, a subdirector, a doctor, a secretary, interpreters, two "writers," a midwife, and sixteen servants.[29] In 1845, the quarantine station of Galați, one of the busiest in the Principalities, had a staff of fifty-four: an administrator (dregător), a deputy administrator, a doctor, a port captain, five commissioners ("of passengers," "of fumigation rooms," "of customs for export," "of customs for import," and "of the shore to receive passengers and goods"), three subcommissioners, a chief secretary, five "writers," an "underservant," twenty servants, three gravediggers, a head sailor, and ten sailors.[30]

Travelers' meals were provided by local contractors, who offered restaurant services. When the quarantine station at Giurgiu was established, a facility of this sort was created. According to official instructions, "the running of this canteen must be well done," and the contractor was to make sure that it was "supplied with all the necessary products, for the supply of food to the quarantine officials and the others passing, having also a kitchen for the making of dishes and always to sell those products according to the prices fixed by the Magistrate in a printed copy."[31]

The introduction of steam navigation on the Danube brought an exponential increase in mobility in the region, obliging the lazarettos to develop

---

28 Scarlat A. Stăncescu, *Din trecutul orașului Giurgiu* (București, 1935), 106–108.
29 Florian Anastasiu, Vasile Anton, and Ștefan Cocioabă, *Monografia județului Brăila* (Brăila, 1971), 71.
30 Bejan et al., *Tezaur*, 95–96. Details in Gheorghe Năstase, Cristina Ionescu, and Rodica Anghel, "Câteva informații despre carantina de la Galați," in Brătescu (ed.), *Din istoria*, 309–310.
31 Stăncescu, *Din trecutul*, 108–110.

their capacity to receive travelers. Most frequented were those of Galați, the port of entry to Moldavia and the hub of steam navigation on the Lower Danube; Brăila, the outlet for the Wallachian cereal trade; Giurgiu, the "port" of Bucharest; and Jupalnic or Orșova, at the entrance to Austrian territory.

Each of these lazarettos was considered illustrative of the level of civilization of its host state. Returning to his native Moldavia after a journey to Istanbul, Teodor Codrescu arrived in Galați in September 1839 on board the steamer *Metternich*. In Codrescu's view, "The position of the quarantine is very appropriate. It consists of 36 rooms, 6 of which are furnished for notable travelers." The good organization was due to *spătar* Iancu Cozoni, the director of the establishment. According to Codrescu, the foreign passengers appreciated the quality of the lazaretto, which "does honor to Moldavia and may rival the foremost in Europe as regards cleanliness and good order."[32] Ruling prince Gheorghe Bibescu also showed his pleasure at the results of an inspection made in the quarantine station at Brăila: after visiting "the passengers' rooms," where there were "24 souls," all content with the conditions they were offered, the prince went "to all the fumigation rooms, to the storerooms, and to those for the airing of goods, [to] the passengers' gardens, the kitchenware store and the place where food is distributed to the passengers, everything he found in the best and most pleasing order. Then he went to the quarantine office, [and] found it in the most pleasing state"[33] (Figure 13).

SANITARY ARREST

Almost all passengers complained about the food and accommodation and also about the medical and policing procedures by which the authorities tried to maintain sanitary and political cleanliness in their respective states. In what follows, I shall present the quarantine experiences of a number of travelers who found themselves guests of the lazarettos of Galați and Jupalnic in the 1830s and 1840s, with details about the organization of quarantine and about the administrative procedures they went through at the beginning of their periods of isolation.

---

32 Gane, *Domnița*, 137–141.
33 Isar, *Sub semnul*.

Figure 13 Bartlett, *Braila—The Lazaretto* (*c*.1840).

In May 1836, shortly after the introduction of steamboat sailings between the Danube and Istanbul, the Prussian playwright and diplomat Karl Otto Ludwig von Arnim (1779–1861) arrived at the quarantine station of Galați. Von Arnim had started a Mediterranean tour at Naples, from where he proceeded to visit the Ionian Islands, mainland Greece, and the Ottoman Empire, and for the journey home, he opted for the Danube route. From Galați, he would make his way via Iași to the Austrian Empire and the German lands. His period of sanitary isolation began with problems, as the lazaretto did not have enough rooms available, and the passengers on the steamboat were asked to wait. Finally, von Arnim arrived on the quarantine quay, a narrow space, delimited on both sides by palisades extending out to the Danube. In his description, he dwells on the accommodation conditions in the "quarantine houses," which he describes as "wooden stables" or rather "rats' nests," furnished with rudimentary wooden beds. There was also a more permanent quarantine house built in stone:

> This has 6 windows in its façade and as many at the back, and 4 doors, 2 on each side. Each door leads to an entrance hall, from where another 2 doors take you straight to the main room and the other to a smaller room beside it, so that the whole house can host four different groups of travelers, each having a day room, a kitchen, and an entrance. However, in this building too there is no other furniture than the abovementioned beds.

Having been moved into one of the "better" rooms, von Arnim enjoyed the luxury of also having a mattress on his bed, "and the uncovered part served as a washing and dressing table. A table, two stools, a few china vases—this was our inventory—thus began our period of quarantine in expectation of freedom."[34]

A similar experience to von Arnim's is described by the Englishwoman Julia Pardoe, the author of numerous volumes of verse, prose, history, and travel narratives. One of her most well-known works is her account of a journey to the Levant in the company of her father, Major Thomas Pardoe. On their way homeward, her group of travelers chose the Danube route, with the result that, at the beginning of October 1836, Pardoe and her suite began their period of isolation in the Austrian lazaretto at Jupalnic (Orşova, see Figure 14):

> After passing through a couple of walled yards, surrounded by warehouses for receiving merchandize, we entered a third enclosure, wherein we were met by the governor and surgeon; who, keeping at a respectful distance, invited us to enter a dark, white-washed, iron-grated cell, in order to have our passports examined.

The officials became more affable when they discovered what class of travelers they were dealing with, and as Miss Pardoe was "only the second lady to

---

34 Karl Otto Ludwig von Arnim, *Flüchtige Bemerkungen eines flüchtigen Reisenden*, vol. 3 (Berlin, 1837), 99–106.

Figure 14 Ludwig Ermini and Friedrich August Wolf, *The Lazaretto of Jupalnic (Orșova)* (*c*.1824).

have been unfortunate enough to come under his keeping," the "governor" (director) offered her the best "cell" (as she considered it to be).[35]

Another gate now opened into the maximum security area of the lazaretto. The travelers found themselves in a courtyard surrounded by a high wall, which offered "the *élite* of the accommodations" available in the establishment: "The cells, like those of a madhouse in Turkey, were built round the four sides of a garden; and each had a small entrance-court, paved with stone." The "cells" themselves, however, were not at all attractive: "windows both barred and grated; walls whitewashed and weather-stained; chairs, tables, and sofa, all of wood, which is a 'non-conductor,' and whitewashed like the walls; were the only objects that met our eyes." The enclosure also had a window, through which it was possible to communicate with "the Restaurateur of the Lazaretto." The travelers ordered dinner, which was served by "a very gaily-dressed, conceited individual, who announced himself

35 Pardoe, *The City*, vol. 2 (London, 1837), 449–450.

to be 'our keeper.'" After "a very bad dinner," the detainees received their bedding, and at sunset the courtyard was locked.[36]

Over the next ten days, the travelers had access to the courtyards of the two quarantine houses in which they were lodged, and which their "keeper" made sure were kept locked. At lunchtime on their first day, they were visited by "the Surgeon of the Lazaretto and the Examining Officer," who was to make an inventory of their possessions: "Each trunk, portmanteau, and basket, was to be unpacked; in short, we were even to declare the contents of our purses!" The travelers unpacked their belongings to be inventoried, and the officials showed particular interest in Turkish items, on which customs duty was to be paid. Pardoe's books and drawings were confiscated to be examined by "the proper officer" (presumably the official responsible for censorship). The gentlemen were required to hand over their weapons, which would be returned to them on their release (an exception being tacitly allowed for a British colonel who refused, as a matter of honor, to surrender his uniform sword). The travelers were visited twice daily by a medical officer, though these visits seemed more a matter of protocol than of serious medical examination. The director also made frequent visits, always maintaining a safe distance with his cane. The detainees were allowed to walk, supervised by their "keeper," to the gate of their friends' courtyard and could receive similar visits in their turn. In the absence of books, there were few pastimes available in the solitude of the cell. The food was relatively unappetizing, so one of the few pleasures available was the company of one's fellow detainees.[37]

The American doctor Valentine Mott, who spent a few days at Jupalnic in 1841, refers to the cells as "prison apartments; for *prison* it was in truth." As his professional reputation was known to the doctor of the lazaretto, Mott received "the very best apartments in the establishment," comprising "a room to myself, one for my companions, also a kitchen, and a large room for the unpacking and airing of our baggage, and the accommodation of our considerate [servant] Henry." The passengers spent the following ten days in relatively decent conditions, with daily visits from the doctor "to inquire after our health, and to know if we were well taken care of, which we in truth

---

36   Ibid., 451–452.
37   Ibid., 452–456.

were; our accommodations now, both in respect to food and lodging, being in every point of view comfortable."[38]

The Scottish doctor John Mason spent three days in the new quarantine station of Galați in 1846. His accommodation there consisted of "one apartment to serve as kitchen, dining-room, and bedroom." It was

furnished with two sofas, made up of a few loose unplaned boards, carelessly nailed together. One of the sofas had something like a mattress, covered with coarse blue drugget. We had also one or two chairs, which, however, were not safe to sit upon, and a very coarse deal table, with a Turkey red table-cover cover, which appeared never to have been intended for the table, as a portion of the table was left uncovered on both ends.

The food he found generally acceptable, "with the exception of the tea, which I should not have known to be tea, unless I had been told so."[39]

As may be observed from these examples, the management of the quarantine establishments favored, as far as possible, travelers of higher social standing, who were accommodated in the cleanest rooms and treated with appropriate respect. The architecture of the lazaretto followed a classic model, and the division into a number of "quarantine houses," depending on the space available and the volume of travelers, facilitated the segregation of travelers who arrived in different groups.

PURIFICATION OF THE BODY AND OF GOODS

(1808–?) mentions how on arrival in the lazaretto the passengers' belongings were unpacked, recorded, "placed on a sort of shelf," and then "fumigated all night." The following morning, "We all had to undress in front of the door, in the open air, and then we were allowed to put on the fumigated clothes and to pack our other things." At the end of the period of quarantine, the travelers would receive a certificate attesting that they had undergone the period of sanitary isolation.[41]

The visual inventorying of naked bodies was not well received by travelers. One famous account of this spectacle of bodies, albeit with only an indirect connection to the Danube, is that of the British admiral Augustus Slade, who, in the lazaretto of Odesa, "admired" the naked bodies of his traveling companions, ranging from Pickwickian corpulence to "well-set herculean strength."[42] The sanitary inspectors were looking for traces of contagion, but also sometimes noted marks of religious identity, as in the case of a group of young Russian Skoptsy, a sect famous for its practice of various forms of castration. After a number of discussions, the authorities in Brăila allowed them to enter the principality, on the grounds that "religious beliefs cannot be used to forbid access to Wallachia."[43]

Interesting as illustrations of the medical conceptions of the period, and of the way in which disease was transmitted, are the procedures for purifying objects, which affected not only the mobility of travelers but also the progress of commercial exchanges. To cleanse goods, the staff of the quarantine station used not only oxidizing products such as chlorine gas or sulfur dioxide (obtained by burning sulfur) but also juniper berries, which were burned as they were rich in volatile oils. Other products used in the fumigation process were vitriol, kitchen salt, and magnesium.[44]

"With regard to their cleansing," according to the Organic Regulations, goods were divided into four "classes." The first comprised products that were not subject to quarantine, such as olive oil, olives, roe, salt fish, and so on. These were imported once the vessels containing them had been washed. The same applied to fruits that could be washed (lemons, citrons, oranges)

---

41 D. Holthaus, *Neue Reisen vollführt in den Jahren 1842–1845* (Bremen, 1846), 105–107.

42 Slade, *Travels*, 309–311.

43 Ștefan Petrescu, "Migrație și carantine în porturile dunărene: controlul documentelor de călătorie în epoca Regulamentelor Organice," *Studii și materiale de istorie modernă* 25 (2012), 97–116.

44 Emil Gheorghiu, "Fumigația, ca mijloc de dezinfecție în carantinele din Țara Românească," in Brătescu (ed.), *Din istoria*, 311–312.

and groceries and other products that could be cleansed once "they are unwrapped and their wrappings burnt or the boxes and vessels containing them are washed with water." Fabrics and textiles were aired for sixteen days, and wool, cotton, and furs for periods that could be up to forty-two days.[45]

Such details are relevant for contemporary visions of epidemics, which were largely based on experience related to plague. Fumigation and washing with water helped remove pathogenic factors. The oilier a foodstuff was, the more inoffensive it appeared, while, conversely, rough textiles were considered very dangerous. We may also note that decontamination practices were relatively standardized in the region.[46]

POLITICAL SURVEILLANCE OF TRAVELERS

As described in the preceding sections, through their function, their spatial structure, and the sanitary procedures applied, lazarettos displayed the full power of the state to subject travelers to various forms of carceral control. Following Michel Foucault, researchers on quarantine have given special attention to the way in which the lazaretto functioned as a panopticon, perfectly equipped to supervise, order, and control the lives of its clients.[47]

Many passengers on the Austrian steamers, especially Western adepts of the miasmatic theory, complained that the aim of the quarantine service was to limit the spread of dangerous ideas more than that of epidemics. The Frenchman Jean-Henri-Abdolonyme Ubicini (1818–1884), who experienced the rigors of the quarantine station at Giurgiu, noted that on entering the establishment he was asked for "a detailed list of my underwear, clothes, and books. I was obliged to say how much small change I had, how many pairs of stockings, trousers, shirts." In the meantime, the lazaretto officials were examining his letters of recommendation and trying to establish if he was a danger to the political health of Wallachia.[48] The Irishman Patrick O'Brien likewise considered that "the quarantine in the Principalities is a polite incarceration of four or five days, during which the police have a very

45   *Regulamentele Organice*, 84, 284.
46   Lukas Engelmann and Christos Lynteris, *Sulphuric Utopias: A History of Maritime Fumigation* (Cambridge, MA, 2020).
47   Michel Foucault, *Discipline and Punish: The Birth of the Prison*, trans. Alan Sheridan (London, 1977); an analysis in Stuart Elden, "Plague, Panopticon, Police," *Surveillance & Society* 1, no. 3 (2003), 240–253.
48   *Anul 1848 în Principatele Române: acte si documente*, vol. 5 (Bucureşti, 1904), 787–792.

necessary facility for making inquiries into your political opinions and your object in visiting the country."[49]

The model applied in the Principalities of Wallachia and Moldavia was far from being original. The Austrians had long been using it successfully, with the imperial lazarettos doubling as centers for political filtering. The Habsburgs were careful to control mobility from the Ottoman Empire, and the entrance points of the Monarchy also served as defenses against the import of ideas and goods that did not conform to official values. Julia Pardoe comments on the Austrian government's "taking an interest in the private affairs, not only of its own subjects, but also in those of strangers," and on the "inconvenient and revolting stretch of power" by which, on arrival at the lazaretto of Jupalnic, each traveler's luggage was examined and inventoried in detail, on the pretext of ensuring that in the event of their decease their family would receive all their belongings. "Certain little circumstances" observed during the process led Pardoe to doubt the intentions of the authorities. In her own case, special attention was given to counting and listing her jewelry and items of clothing. As mentioned above, her books and drawings were confiscated to be examined for censorship purposes. Such examples confirm the studies that situate the lazarettos among the institutions that applied authoritarian and antidemocratic policies. These establishments were spaces and structures of power in which the state made use of various sanitary procedures for the political control of mobility. The lazarettos were thus, at the same time and in equal measure, points of sanitary control and instruments of national security and state-building.[50]

THE LAZARETTO AS A CONTACT ZONE

Apart from these unpleasant aspects—or perhaps in fact because of them—the period of quarantine was a memorable one for most travelers. They spent their period of sanitary arrest in a group that was often ethnically, religiously, or socially heterogenous, in a situation that required them to establish connections with other passengers in order to overcome the privations more easily, thus contributing to the formation of a spirit of comradeship among

---

49 O'Brien, *Journal*, 13. Also see Petrescu, "Migraţie şi carantine," 98.
50 Chircop and Martínez, "Introduction: Mediterranean Quarantine Disclosed—Space, Identity and Power," in *Mediterranean Quarantines*, 1–14.

those detained together. The lazaretto was, by definition, a global site, well connected to the world, whose space and time varied according to events and information arriving across land and sea. It was a contact zone for travelers from different worlds, offering a fascinating transnational dimension to their experience.

For travelers obliged to remain in the lazaretto for a number of days, socializing with their comrades in detention was one of the few relaxing activities available. The Prussian von Arnim notes the familiarity of persons and discussions, for example, with a Moldavian (dressed in no more than a dressing gown, "for who would be embarrassed in a quarantine?") who lent him a mattress "that I was allowed to keep throughout my imprisonment." The day's program consisted of reading and writing in the morning, followed by dressing and grooming at lunchtime: then "one goes for a walk, eats lunch, then works again and so, after another walk, the time comes round for tea and supper." The playwright repeatedly socialized over tea or supper with the Cantacuzino family, who were also in the establishment.[51]

The French merchant Jean-Baptiste Morot (1797–?), who was detained for twelve days at Jupalnic in September 1840, gives a detailed account of his experience in the Austrian lazaretto. His group agreed that an Ottoman subject and his wife, together with whom they had traveled on the steamer, should join them in the quarantine house they had been allocated: "When dinner time came, we were allowed to eat together. For lack of a large room (there were more than twelve of us), they put the table in one of the yards, the most spacious. Lunch was very merry, in spite of the diversity of languages, because we were all of different nationalities; only appetite was foreign to no one."[52]

The British captain James John Best also mentions the camaraderie formed with his comrades in suffering, a group that also included an American traveler. "This was my first intimate acquaintance with an American, and I must pay him the compliment of saying I shall be extremely happy to meet him again in any part of the world, and if it should be our fate to have to undergo another period of imprisonment together, to have him as my companion."[53]

---

51 Arnim, *Flüchtige Bemerkungen*, 102–107.
52 Jean-Baptiste Morot, *Journal de voyage. Paris à Jérusalem. 1839 et 1840*, 2nd ed. (Paris, 1873), 334–340.
53 Best, *Excursions*, 323.

Andersen spent his ten days of quarantine in 1841 in the company of Francis W. Ainsworth. Both described their time in the lazaretto of Jupalnic. The wider group included some Austrian officers, an Armenian priest accompanying his bishop's young nephews, "several French leech-dealers," some Bulgarian women with their children, and two Wallachian musicians. The ten days passed slowly, and Andersen remarks that afterwards, just as a traveler continues to feel seasick for some time after disembarking, he remained with "a feeling of the quarantine."[54]

The lazaretto was also a space of familiarization with local characters. Detainees frequently recall the director, responsible for administrative aspects, and the doctor coming on his rounds. However, it was the warden in charge of each house who both kept watch on the behavior of the detainees, making sure they followed the isolation regulations, and checked that they lacked nothing they were entitled to. Their interaction with the warden-cum-servant was often amusing, and travelers made use of this figure to underline the tragicomic situation in which they found themselves. Ion Codru-Drăgușanu recalls the good-humored servant, who "both with jokes, and even more with his Banat [regional] speech, entertained us. Such a person is a treasure in quarantine, which resembles the harshest arrest."[55] Best recommends "my friend Lasar [Lazăr], our guardian" as "a most willing, excellent fellow, and a capital servant,"[56] and Anderson recalls "Johan" (Ioan), a veteran of the Battle of the Nations at Leipzig (1813), who was employed at the establishment together with his wife, who worked as a washerwoman.[57]

Patrick O'Brien, too, recalled the warden who guarded and served him in the lazaretto of Brăila. His serving at table, in particular, left much to be desired:

> At dinner hour, for example, he appeared with a basin in one hand and an earthen dish in the other. In the basin was soup, and on the dish boiled meat or pilaff, or both together; and about his person he carried the rest of the dinner, and at times some small article which he did not find room

54 Andersen, *A Poet's Bazaar*, 166–178; Francis W. Ainsworth, "Herr Andersen," *Literary Gazette* 1551 (October 10, 1846), 877; Andersen, *The Story of My Life* (Boston, 1871), 168–169.
55 Codru-Drăgușanu, *Călătoriile*, 19–24.
56 Best, *Excursions*, 323.
57 Andersen, *A Poet's Bazaar*, 3, 170, 176.

for in his pockets he held between his teeth. After he had laid the basin and the dish on the table, he drew forth a little plate, with a very small iron fork, a spoon of the same metal, and a rusty knife. Off the same little plate I ate the soup, slowly and painfully, as well as the pilaff and meat, or whatever else there might be. I made no attempt at having my *couvert* changed with each dish; for on the first day, when I asked the guardiano to clean the plate after I had eaten my soup, I saw that he was preparing to do so with a cloth which he drew out of his pocket.

The bill that O'Brien had to pay at the end of his four days of quarantine seemed somewhat excessive: 10 francs per day for food, plus rent for his cell and "the wages of my intelligent guardian."[58]

As a continuation of experiences on board the steamboats, the lazarettos were spaces not only of segregation and exclusion but also of intense socialization within the small groups that had to share a quarantine house. The lazaretto's combination of restricted space and abundant free time helped cement comradeship among its inmates and familiarized them with aspects of private life that were little visible in other social settings. It thus became another contact zone, in Mary Louise Pratt's sense of "social spaces where cultures meet, clash, and grapple with each other, often in contexts of highly asymmetrical relations of power."[59]

A Visit beyond the Cordon Sanitaire

Baron Frederick John Monson (1809–1841) traveled frequently, not only for tourism but also in the interests of his generally fragile health. One of his journeys took place in 1839, when he visited a large part of Europe, finally arriving in the Romanian space after traveling by steamer from Budapest. On the afternoon of October 2, 1839, while he was in Orşova, Monson set out on one of the most interesting excursions travelers in the region could make: a visit to the island of Ada Kaleh, also known as New Orşova.

Ada Kaleh was as fascinating in the nineteenth century as it is today, submerged as it now is under the waters of Iron Gates Dam I. The position of

---

58 O'Brien, *Journal*, 14–17.
59 Mary Louise Pratt, "Arts of the Contact Zone," *Profession* (1991), 33–40.

the island in a picturesque natural setting[60]—in a space that, as we shall be reminded in Chapter 6, was seen as the gateway between East and West—attracted the instinctive attention of tourists. The fortress on the island conferred on it the role of a symbolic space for the situation of Ottoman power in Europe. Dimitrie Bolintineanu had few words of appreciation for this settlement inhabited by Turks, Serbs, and Romanians: "The streets are dirty; the houses have a miserable physiognomy; the walls of the fortress are decayed and rusty cannons sit on the bastions. The most unhealthy damp, coupled with the deepest squalor, reign in this place."[61] However, it was this very degradation, this anchoring in squalor, that was attractive.

Visiting Ada Kaleh was a simple matter for travelers on the right bank of the Danube. However, from Austrian Orşova or Romanian Turnu Severin, an excursion to the island was impossible without having, at return, to go through quarantine procedures. The solution offered by the authorities was that the visit could be made under conditions of sanitary and customs supervision, with officials employed to accompany the group on the island and ensure that the rules of antiepidemic protection were respected and that visitors did not bring back contraband products from the Ottoman Empire, especially tobacco, which was heavily taxed by the Habsburg authorities.

Monson's account is representative of the course of such an excursion. The baron and his fellow visitors, accompanied by two Austrian officials, had arranged to pay a visit to Pasha Mahmud Bessim, commandant of the fort of New Orşova, and to see the island. Upon disembarking, the visitors were met by the pasha's dragoman and two Turkish soldiers. Monson's description of one of the soldiers, "a thin, tall, aged, scarecrow looking fellow, with a dress faded and ragged, and a musket which seemed more for ornament than use," serves to confirm his expectations of decay. The visit to the pasha's residence was a veritable display of protocol in a time of sanitary troubles: "In the middle of the room was a mat, with a row of wooden chairs prepared for our reception, on which we were informed we might seat ourselves, and partake of such refreshments as would be offered to us. We were directed to keep on our hats during the interview, such being Turkish etiquette." After the

---

60 Skene, *Wayfaring Sketches*, 288–289.
61 Bolintineanu, *Călătorii pe Dunăre şi în Bulgaria*, 5.

initial pleasantries, "The attendants now advanced towards us, and kneeling down presented each of us with an amber-mouthed pipe." After "a few silent whiffs" and further pleasantries, the dragoman, taking care not to touch the guests even with his robe, served them "some unknown but delicious sweet-meats, and some glasses of water," which were followed by what Monson considered "some of the most delicious coffee I ever tasted, certainly far superior to any I have met with in England or elsewhere." The audience lasted some twenty minutes, at the end of which "we now rose, made our salaam by laying the right hand upon the heart and bending the body forwards, and quitted 'the presence,' after a warning from the quarantine officer not to step off the mat upon the floor of the room; a precaution perfectly absurd, after having sat on their chairs, and held the cherry-stick pipes in our hands, &c." Being responsible for protocol, the customs official gave the dragoman a small tip. There followed a visit to the town, which had the usual appearance of an Oriental settlement, with "narrow streets" and houses in the form of "wretched huts of wood or mud." As for the inhabitants, "The people whom we met seemed to have quite as great a dread of coming in contact with us as we had of them, running away where that was possible, or squeezing themselves into a corner to give us forty times more room to pass than was necessary." The shops had nothing interesting to offer—fortunately, as in any case the visitors were forbidden to buy any potentially contaminated items[62] (Figures 15 and 16).

Numerous other travelers both before and after Monson made the same excursion into the Ottoman world. The Prussian officer Helmuth von Moltke (1800–1991) visited the commandant of the fort, Osman Pasha, in October 1835. The pasha sat in an armchair and his guests on cane seats, positioned so as not to touch the mat spread on the floor (which might transmit plague). The festivity of serving pipes and coffee followed with the same precaution. The coffee was served in cups placed on little silver trays. A servant disinfected them before handing them to the guests.[63] Johann George Kohl also insisted on meeting the pasha, who proved, as on all the other occasions mentioned, to be very well informed.[64]

---

62 Frederick John Monson, *Journal of a Tour in Germany* (London, 1840), 147–157.
63 Helmuth von Moltke, "Tagebuch der Reise nach Konstantinopel," in *Gesammelte Schriften und Denkwürdigkeiten des General-Feldmarschalls Grafen Helmuth von Moltke*, vol. 1 (Berlin, 1892), 128–129.
64 Kohl, *Austria*, 274–276.

Figure 15 John Richard Coke Smyth and John Frederick Lewis, *New Orsova;
Exterior of the Pasha's Residence* (1838).

Figure 16 Smyth and Lewis, *Interior of the Residence of the Pasha,
New Orsova* (1838).

All these accounts are illustrative of the manner in which the image of the Orient is constructed from clichés and stereotypes, which in fact are representative of the author's own identity.[65] What Ada Kaleh had in addition was the unique symbolic status of an island with a special place in the mental cartography of Europe, an enclave lost in space and time, between East and West, between past and present. As a sight to be seen, the island of Ada Kaleh was (and is) a monument, a site of living memory, an allegory of the past and of the eternal.[66] If one was traveling in the region, it was most definitely worth a visit.

COMMERCE IN TIMES OF ANTIEPIDEMIC PREVENTION

The play of commercial exchange with territories beyond the cordon sanitaire and the asymmetrical situation of trade in a time of epidemiological threat could be seen in the "parlatoria" of the Danube, enclosures that served to ensure physical distancing between traders. As described by Monson, "The Parlatorium [of Orşova] is a long oblong shed, open all round, but roofed in. It is divided into three parts by wooden palings, and in the center is another place partitioned, with a table in it for the quarantine and custom-house officers, who are on these occasions attended by an armed guard." These officials supervised exchanges from both the sanitary and the fiscal points of view. Goods were bought and sold through their intermediary after being either washed in vinegar or fumigated, according to the nature of the produce.[67]
Kohl adds:

> The Austrians are at liberty to sell every thing to the Turks, but are allowed to purchase from the latter only such merchandise as are not deemed liable to infection, such as corn, fruit, meat, wood and the like. As soon as they have agreed on the price, if it is the Turk who has to pay, he throws his money into a vessel filled with water, whence it is the Austrian's business to fish it up again. Austrian health officers and sentinels meanwhile are walking up and down in the intermediate inclosure to see that the rules and regulations are strictly complied with.[68]

---

65 Said, *Orientalism*.
66 Király, *"Die Donau ist die Form,"* 356–362.
67 Monson, *Journal*, 145–146.
68 Kohl, *Austria*, 278.

As Orşova was an important point on the route between Vienna and Istanbul, and passengers going downstream had to wait here for the transfer of their luggage through the Danube Gorge, a visit to the "parlatorium" was one of the usual excursions made by travelers, who thus became familiarized with the rigors of quarantine. The process is also mentioned in Murray's guidebook, starting from the first edition, which emphasizes the "amusing sight" of "the process of bargaining thus carried on by 3 parties at the distance of several yards from each other, attended by the vociferation and gesticulation inseparable from such business," all made more interesting by the prevention measures carefully supervised by the Austrian officials.[69]

By Way of Conclusions: Contagionists vs. Miasmatics

The region of the Lower Danube was the place of convergence of contradictory visions regarding the role of quarantine in preventing the spread of plague and cholera epidemics, which in fact corresponded to a more general dispute about the propagation of disease. According to the "contagionist" theory, which found more favor within the medical community in Eastern Europe, epidemics were transmitted through direct contact, through touch, through *contagion*. The quarantine station was thus a place for the isolation and "cleansing" of those coming from zones of epidemiological uncertainty, and above all for waiting for the time to pass in which the disease might manifest itself. Its role was to set up a spatial but also temporal barrier separating the "clean" world from the possibly "contaminated" one.

For adepts of the "miasmatic" theory, which was more prevalent in the medical community of the West and in Western society in general, the culprit in the transmission of disease was "miasma" or fetid air, which made its presence felt in dirty accommodation. Starting from the observation that epidemics were more numerous and more virulent in the "uncivilized" territories of the Orient, the miasmatics linked disease to hygiene, and especially public hygiene. Thus, for most Westerners familiarized with Oriental conditions, the issue was not how to control mobility between the two worlds but how to attempt to "purify" the miasma—in other words, how to eradicate

---

69   *A Handbook* (1837), 385.

the deeper cause of sickness.[70] Lazarettos, they believed, most of which were spaces of precarious hygiene, were themselves responsible for the spread of diseases. As underlined in a recent article, the distinction between the two visions is not an absolute but rather a fluid one,[71] and many medical practitioners accepted an intermediary version between the contagionist and miasmatic extremes.

Lazarettos and periods of quarantine were an expression of the contagionism that had been so common in past centuries in Europe, hence their contestation by many doctors and by travelers who believed in the progress of medical science. Quarantine had operated efficiently in past centuries, in times of much less mobility. Now it was time to move on, toward solving problems. Merely imposing sanitary barriers and waiting for the test of time was not enough. The time had come for the character of epidemics to be studied in detail. Were they contagious or infectious? It was then vital for the etiology of diseases to be researched and for action to be taken with maximum administrative efficiency in order to be prepared for epidemics. Equally imperative was a better management of waiting times, in a world that had less and less leisure to stay still.

The introduction of steam navigation on the Danube, as on the seas of southern Europe, put additional pressure on the quarantine system. The growth in international mobility and economic exchanges was incompatible with the classic system of medieval and premodern quarantine. In spite of the efforts of the states in the region—which invested in the equipping of lazarettos and in increasing their capacity to receive passengers and goods— the imposition of long periods of quarantine was not viable in the face of the growing demand on the mobility market. The pressure of the shipping companies, which had considerable support at the political level, contributed to a gradual reduction in waiting times, as the Sublime Porte established its own quarantine system and modern technology contributed to the rapid transmission of information about the spread of epidemics.

The management of information was thus vital, and it was this that determined the exact duration of the period of cleansing, depending on the

---

70   TNA, FO 195/285, fol. 2–4 (Charles Cunningham to Henry Wellesley, Galaţi, February 18, 1847).
71   Jon Arrizabalaga and Juan Carlos García-Reyes, "Contagion Controversies on Cholera and Yellow Fever in Mid Nineteenth-Century Spain: The Case of Nicasio Landa," in Chircop and Martinez (eds.), *Mediterranean Quarantines*, 170–195.

sanitary realities of the Ottoman Empire. The tendency was one of continual reduction in the period before the Crimean War, with variations associated with epidemics in the region. In periods when there were no active outbreaks, travelers could enter the Habsburg Empire after just one day, as it was considered that they had begun their period of quarantine as soon as they left Istanbul.[72]

Political aspects contributed to complicating the situation. Against the background of the cholera epidemic of 1848, together with the revolutionary movements in the region, Russia increased its control of the Moldavian and Wallachian quarantine stations, where, during the period of military occupation of the Principalities, Russian agents had a surveillance role.[73] The ruling princes appointed in 1849, Grigore Alexandru Ghica and Barbu Știrbei, tried, at the urging of the Porte as well as in their own interests, to limit the interference of Russia in the quarantine services of the Principalities.[74] The defeat of Russia in the Crimean War came as a heavy blow to the quarantine system, which was first reformed[75] and then eliminated, only to be rethought and reimposed in a form better suited to the age of full economic freedom inaugurated by the Peace Treaty of Paris (1856).

---

72 C. F. Fynes Clinton, "A Few Pages from My Journal in Greece, Turkey, and on the Danube," *Bentley's Miscellany* 16 (1844), 340.

73 TNA, FO 195/332, fol. 18 (R. G. Colquhoun to Stratford Canning, Bucharest, January 19, 1849); FO 195/349, fol. 125 (Cunningham to Canning, Galați, May 18, 1850).

74 *Mărturii istorice*, 163–164, 610–611.

75 Trăușan-Matu, "Doctorul Nicolae Gussi și istoria carantinei în Țara Românească," *Jurnal Medical Brașovean* 1 (2020), 138–139.

CHAPTER 6

Nature and Technology at the Iron Gates

THROUGH "THE GAPING JAWS OF SOME INFERNAL MONSTER"

The Irishman Michael J. Quin was one of the first Western authors whose account of a journey on the Danube became a bestseller. As mentioned in Chapter 4, Quin embarked on board a DDSG steamer in Budapest in September 1834, on his way from Paris to Istanbul. After several days sailing across the Hungarian plains, the approach of the Iron Gates gorge brought a pleasing change of landscape. The passengers arrived at Moldova Nouă on September 29, but the exceptionally low level of the river ("not six inches of water, nor even three") prevented their being transferred by wherry to Orşova as planned. The steamer's cargo was to be transported overland, while the travelers themselves had to choose between continuing down the river in the "rough flat-bottomed boat" of a local fisherman and going on foot "over horrible mountain roads." The DDSG agent, an Italian, assured them that if they went with the fisherman, they would arrive in Orşova in eight hours at the most.[1]

Moldova Nouă was rapidly developing, and the circulation of the steamer seemed to be stirring the entrepreneurial spirit of the locals. Quin visited the village and, from a nearby hill, admired the landscape, pondering on the processes by which the Danube had come to break through the barrier of

---

[1]   Quin, *A Steam Voyage*, 70–72.

the Carpathian Mountains. All around was picturesque and offered Quin a pleasant day of relaxation. He spent the night back on the steamer, as there was no inn in the village.[2]

On September 30, there followed an unforgettable experience. The small group of passengers on the fishing wherry was delighted by the natural surroundings through which they were passing and at the same time horrified at the apparent indolence of the crew, headed by a septuagenarian "patron" at the helm, who had serious problems with his eyesight. The passengers found the landscape became more and more fascinating as they advanced downstream. The grandeur of nature and the ephemerality of human existence passed through their minds as they were carried by the current on a wild race through space and time. On a majestic crag, they could see the ruins of the fortress of Golubac (Figure 17), once the refuge of some Wallachian outlaws, a mysterious place that inspired the most romantic tales. Then the sounds of nature were rudely interrupted by a series of explosions that sounded like artillery fire. Their true nature became apparent when the passengers saw teams of workers busy widening a road cut into the rocky mountainside on the left bank. For Quin, the explosions "spoke of enterprise and industry well applied, and were the harbingers of national prosperity, civilization and happiness." There followed a number of caves, once the shelters of pirates and brigands, some of them still occupied by fishermen. In this splendid landscape, even the mountains themselves displayed artistic talent:

> Occasionally masses of rock appeared above our heads, depending for support on rude pillars, in which capitals wrought by the hand of nature might be descried. One immense buttress rose in the shape of a round tower, near the top of which a large cavern was visible, accessible by a gateway naturally arched in the Gothic style.[3]

Only the obvious negligence of the sailors spoiled the beauty of the journey. The boat was almost wrecked on some rocks in the middle of the river, reminding everyone of the perils of the route. The group then approached

---

2   Ibid., 74–78.
3   Ibid., 79–84.

Figure 17 Bartlett, *Ruins of the Castle of Golumbacz* (*c.*1840).

the stretch of rapids, where the bed of the Danube was completely covered with rocks that showed their threatening teeth on the surface of the water. Danger seemed imminent, announced from a distance by "the hoarse murmur of the waters." The boat was caught in the current, and the banks seemed ever closer and more threatening. The bottom of the boat rubbed against the rocks, as there was insufficient water to enable it to float. The helmsman said he had never seen the water level so low. The play of images continued, with the rocks disguising themselves in the most varied forms: a water mill, a monk, a lion, even "the ruins of a cathedral, with its towers and ivied walls, and Gothic windows and gates." The imagination was free to rove, but evening was rapidly falling and a delayed departure from Moldova Nouă meant that it would be impossible to reach Orşova on time. The passengers instructed the helmsman to let them disembark on the left bank and headed for the village of Sviniţa. In the village inn, where the "Moldavian adventurer" Teodor Burada demonstrated his talents as a "magician" (see Chapter 4), Quin was surprised to make the acquaintance of an Englishman, George

Dewar, who was employed to operate the diving bell used by the Hungarian engineers who were trying to improve navigability in the Danube Gorge.[4]

The travelers resumed their journey in the same boat the following morning. The carousel of forms continued, as the surroundings changed rapidly, always in surprising ways. The workers on the left bank had been busy since dawn, and the noise of explosions, together with "the mallet and punch, the pickaxe and chisel," drowned out the sounds of nature. The road was starting to take shape, including, where necessary, "bridges or terraces [...] erected in a solid and, at the same time, an ornamental style, which reminded me of old Roman enterprise" (Figures 18 and 19). Having gone ashore to examine the works more closely, Quin was shown round by an Austrian officer, who took him to "the auberge of the colony"—a natural cave, where strange rock formations towered over the human beings working there. Indeed, he writes,

> wherever I looked around me, it appeared as if I had found a mystic portion of the globe, which, like the face of Satan, "deep scars of thunder had intrenched;" where Chaos still held her reign, and none save the Titans of elder time could hope to dwell in security. But my terrors were reproved by some young saplings which burst forth from amidst the rocks, spreading their graceful branches in the air. Here and there a wild flower, too, displayed its blue or coral bell; the bee murmured quietly along, the sparrow twittered, the yellow butterfly wandered about, and the spider floated by in his gossamer balloon.

Continuing on their way, the travelers visited "Veterani's cave," the famous refuge of the Austrian general Friedrich von Veterani (1650–1695), who in 1691 resisted a long Ottoman siege in this natural fortress. The explosions outside seemed a continuation of the confrontations between empires, and indeed, symbolically speaking, so they were.[5]

At Orșova (Figure 20), the passengers were met by the DDSG agent, one Popovicz, together with a much more illustrious host, Count Széchenyi István. Accommodated in a "very respectable" hotel, Quin had some time to relax while the steamer, anchored off the Serbian village of Kladovo, waited

---

4   Ibid., 85–97.
5   Ibid., 98–115.

Figure 18 Bartlett, *Entrance to the Defile of Kasan* (*c*.1840).

Figure 19 Bartlett, *The Kasan Pass, with the Modern and Roman Roads* (*c*.1840).

for the arrival of the goods, including some carriages, that had been sent over-land from Moldova Nouă. He visited the mart in Orşova, where business was carried on between merchants from the two sides of the river in conditions of physical distancing similar to those described in Chapter 5. He then dined with Széchenyi and some of the count's collaborators in the project of regu-larizing transport through the Danube Gorge. As well as the works on the river bed, the plan included the creation of a carriageway along the left bank. The considerable cost of the works was supported by the Austrian govern-ment. As the initiator and overseer of the works, Széchenyi was fully aware of the great advantages to the imperial authorities, and indeed his native Hungary, that would result from the opening of a commercial highway along the Danube toward the Black Sea. Quin understood equally well the wider effects of the daring technical enterprise coordinated by the Hungarian nobleman: "The steam navigation of the Danube will also be a most power-ful instrument of civilization; for it is quite true that steam and civilization are daily becoming almost convertible terms. Wherever one of these is found, the other cannot be far distant." The taming of nature and the opening up of the Danube to international commerce announced a new age of freedom for the countries along its banks. For the lands of Central Europe, like Hungary, steam navigation was a veritable declaration of independence, not only from the authority of nature but also from that of the Habsburg Empire.[6]

On October 3, 1834, Quin accompanied Széchenyi in the latter's phaeton on the road to Schela Cladovei. They soon arrived at the stretch of the river known as the "Iron Door":

> It is a series of rapids so called from the extreme difficulty of passing them, and also probably from the almost impenetrable nature and ferru-ginous colour of the rocks, which form the entire bed of the river to the distance of nearly three miles. These rocks, though so long washed by the torrent are still as rough as when the river first found or forced its way amongst them. They are in large masses, tumbled about in every sort of shape and position, and now that they were completely exposed to view, in consequence of the depression of the river, they looked terrific; the gaping jaws, as it were, of some infernal monster. When the Danube is at

---

6 Ibid., 116–125.

Figure 20 Ermini and Wolf, *Cerna's Mouth at Orşova* (*c.*1824).

its ordinary height, replenished by its usual tributaries, the roar of its waters in hurrying through the "Iron Door," is borne on the winds for many miles around, like the sound of continued peals of thunder.[7]

The rocks divided the course of the Danube into three channels, of which that on the Serbian side was navigable for vessels with low draft. The current was extremely strong, which meant a further complication for commercial traffic. The passengers admired Trajan's Tablet, the testimony in stone to another engineering work—a Roman road—of great strategic value. Arriving at Schela Cladovei and finding that the goods were still on their way from Moldova Nouă, the travelers, accompanied by a quarantine inspector, crossed the river to visit the Serbian town of Kladovo (Figure 21) on the opposite bank and then, returning to the Wallachian shore, took advantage of the low water level to explore the presumed site of the bridge of Apollodorus of Damascus. Inscribed in the rock, the history of the confrontation between nature and civilization was now entering a new episode, in which some of

7   Ibid., 144.

Figure 21 Bartlett, *Village of Gladovo* (c.1840).

the most able engineers of their time were building symbolic bridges over or along the rocky Danube Gorge.[8]

Having finally embarked on the *Argo*, the passengers left Kladovo on October 6, 1834, more than a week after their arrival in Moldova Nouă. Clearly, the transfer through the Danube Gorge would have to be made faster, easier, and more comfortable in future. According to the plans of the DDSG, on the new route, travelers would get from Vienna to Istanbul in around eight days, rather than about three weeks required to make the journey overland. The Danube route would, Quin remarks, "exhibit, therefore, one of the most important triumphs over time which the steam-engine has yet accomplished."[9]

The route brought considerable advantages to Hungary and the other territories along the Danube, as these lands on the periphery of Europe could now enjoy the advantages of modern civilization. Mobility on the continent would grow accordingly, permitting political elites, merchants, and tourists to visit the principal cities of Europe in two or three months—something

---

8   Ibid., 145–151.
9   Ibid., 152.

that could not previously have been achieved in less than a year. With the challenge of the new vessels sailing on the great river, the Danube Gorge promised to become a veritable shortcut through space-time, a wormhole or a chronotope in the transport infrastructure of Europe. The region was now attractive not only for its role as a connector between river sectors that had previously been economically separate but also for the way in which it had become a zone of meeting between two apparently antagonistic realities: nature at its most perfect, and technological civilization at its most modern.

Starting from Quin's presentation, this chapter aims to present the region of the Danube Gorge as a special space, a heterotopia, that mobilized the resources of the DDSG and aroused the interest of passengers on the Austrian steamers. On the geological and natural level, the meeting of the river and the mountain created a natural setting as varied and attractive as it was difficult to "tame" from a technical point of view. Numerous relics of Roman material civilization and the ruins of medieval fortifications bore witness to the turbulent history of efforts to make the region more accessible or, on the contrary, impenetrable to human mobility. Geopolitically speaking, the gorge was a contact zone of the Habsburg monarchy and the Ottoman Empire, with sectors of the river belonging to the Principalities of Serbia and Wallachia, autonomous states that were trying to promote their own interests, including in relation to navigation on the Danube.

For all these reasons, the Danube Gorge was a transition point, a threshold, a gateway,[10] a fact that is even better communicated in the name by which the area was (and still is) known: the Iron Gates. Variously perceived by travelers as a symbolic gate between Christianity and Islam, between East and West, between barbarism and civilization, the Danube Gorge was opened to steamboat navigation and the admiration of passengers, who gave it their full attention, discovering in it, or charging it with the aesthetic values specific to the Romantic period. Travel through the region became a guided tour through an open-air museum in which the passengers could admire the grandeur of nature and perfect landscapes and could cultivate their passion for ruins or for technological inventions, all included in the price of the ticket and ready to be explored in a complete experience.

---

10 Király, "*Die Donau ist die Form,*" 345–388.

Figure 22 *Map of the Danube Gorge (c.1844).*

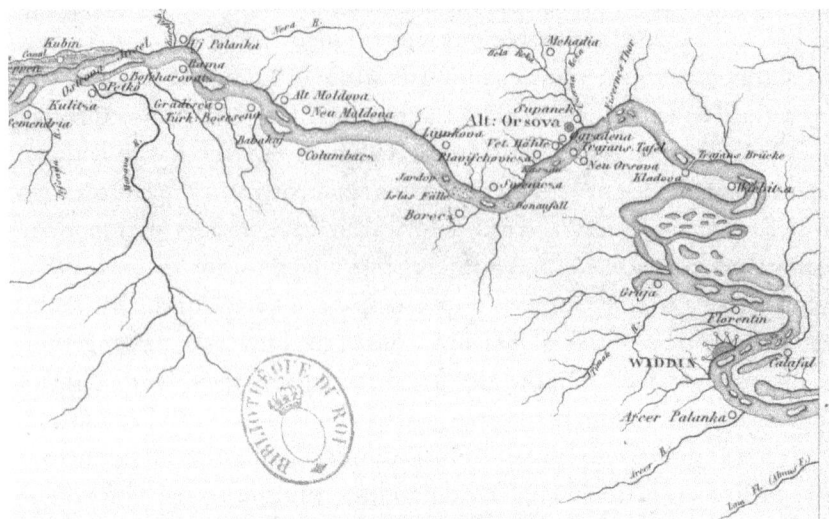

In the following pages, I shall present the principal stages in the invention of the region of the Danube Gorge as a special space where nature and technology met, with direct reference to the contribution of engineers to the "conquest" of the natural environment and of the passengers on the DDSG steamers to its endowment with the traits of the Romantic period.[11] References to the technical projects to facilitate navigation and the political arrangements that favored such initiatives will be followed by discussion of the aesthetic valorization of the region, part of the way in which the authors felt the need to give metaphorical interpretations to their experiences along the river.[12] Not least, I shall present in detail the logistical aspects that permitted the circulation of passengers through the gorge (Figure 22), together with some of their descriptions of this rite of passage through the corridor between worlds.

"A Gigantic Project" of Disciplining Nature

The Danube Gorge stretches over some 150 km, between Baziaş and Drobeta-Turnu Severin. The region includes a number of obstacles in the

---

11   Marian Popa, *Călătoriile epocii romantice* (Bucureşti, 1972), 220–239.
12   Burroughs, "Travel Writing and Rivers," 331.

way of navigation, to which I shall return in this section. From the earliest historical times, the gorge was so difficult to navigate that the geographers of antiquity, such as Strabo, considered that it marked the separation between two distinct rivers, the Danube and the Ister.[13] The distinction persisted on the geographical, economic, and mental levels, as there were relatively few economic connections between the Danube upstream and the Ister downstream. History records a number of projects aimed at regularizing the course of the river, but these proved too difficult technically or too costly to become reality. Of course, the gorge was circulated by the light boats of locals familiar with the characteristics of the area, but navigation with medium- or large-tonnage vessels was either impossible or not commercially viable.

The introduction of steam navigation on the Danube in 1830 created revolutionary opportunities with regard to the economic unification of the disconnected sections of the river. Plans were soon revived for the extension of Habsburg commercial penetration beyond the Iron Gates, toward not only the Lower Danube but also the Russian Empire's Black Sea provinces, the Ottoman Empire, the Caucasus, and India. Such projects had first been aired a century earlier, but the period of political and military turbulence that had intervened had made it impossible to profit fully from the opening of the Black Sea ports for international trade. The Peace Treaty of Adrianople with its economic provisions opened new opportunities for the business community in the Austrian Empire, and here the business plans of the DDSG and the government in Vienna seemed to line up.

The removal of the physical obstacles in the way of navigation through the Danube Gorge was a daring process of conquest, taming, or disciplining of nature.[14] The project began in the 1830s and continued, in a number of phases, throughout the nineteenth century, culminating in a festive inauguration in 1896, the symbolic year when the Hungarian monarchy celebrated a thousand years of existence. The modeling of nature through the activity of engineers proceeded at the same time as the fixing of the political map through the formation of the nation states of the region. In this chapter, I shall refer only to the first phase of regularization, that of the visionaries who

---

13   *A Handbook* (1857), 532.
14   L. Iancovici, "Un proiect pentru înlesnirea navigației pe Dunăre la Porțile de Fier din anul 1833," *Analele Universității București. Istorie* 23, no. 1 (1974), 189–191.

made use of imagination and daring as much as of technology in their conquest of nature. Two figures had a decisive role in the success of the initiative of conquering nature: Count Széchenyi and the engineer Pál Vásárhelyi.

Széchenyi was aware that it was a "gigantic project" that necessitated immense material resources and energy to match. The count was well endowed with both, and his investments, both material and personal, proved decisive for the success of the project. The firstborn of an influential noble family, Széchenyi had begun his public career holding junior posts in the Habsburg army at the time of the Napoleonic wars, before going on to try out his diplomatic abilities as imperial representative to the court of Naples. He traveled frequently throughout Europe and became a vocal admirer of British civilization, which he considered the crowning of the relation between material property and cultural achievement. In 1826, he became a member of the Hungarian Diet, thus beginning an illustrious career that was to cause the following two decades in the history of Hungary to be named "the age of Széchenyi." His speech in Hungarian in a Diet whose official language was still Latin surprised many, as did his donation of a large sum of money that permitted the creation of the Hungarian Academy. Széchenyi showed a real interest in improving the social situation of the peasants on his estates, supporting the development of their material condition by means of reforms aimed at bringing the Hungarian economy out of feudal backwardness. In what one historian has termed "aristocratic entrepreneurship," the count was equally interested in equipping the peasants with the tools of modern agriculture, in attracting them to the cultivation of silkworms, in setting up the system of credit and the first Hungarian banks, and in developing local industry. In Budapest, one of his most important projects, the Széchenyi chain bridge, which in 1849 joined the two banks of the Danube, is still to be seen. It was a great success not only technically but also politically and socially, as its construction necessitated the elaboration of a modern legislative framework that annulled some of the privileges of the Hungarian nobility, who had been exempt from the toll to cross the old bridge of boats in what today is the capital of Hungary.[15]

As a Hungarian patriot, Széchenyi was a great supporter of the building of another sort of bridge, a symbolic one over or through the Carpathian

---

15 Nicolson, "Count Istvan Széchenyi," 163–180.

Mountains, to unite the Danube, whose course was fractured in many places, especially at the Iron Gates. When in 1833 he was appointed royal commissioner for the development of Danube navigation, Széchenyi poured himself into the project of regularizing the Danube Gorge and pushing imperial interests toward the Lower Danube. This was an important political interest of the government in Vienna, which Széchenyi also supported because it was vital for the development of the Hungarian territories of the empire. Historians still discuss his true loyalty, which, more than to either of the two parties that in 1848–1849 would find themselves in conflict, was to progress.[16]

THE CONQUEST OF NATURE

In recent decades, environmental historians have given increasing attention to one of the most important revolutions of the modern age: the "conquest of nature," especially through massive works to channel rivers—in other words, to "domesticate" them and to use their resources in a "rational" way, to the greatest possible benefit of human communities along their banks. In an influential study, Mark Cioc has written an ecobiography of the Rhine, a river "civilized" by the great hydraulic works coordinated by Johann Gottfried Tulla (1770–1828), the engineer who "tamed" the great German river.[17] In another influential work, *The Conquest of Nature*, David Blackbourn amply demonstrates how, in a new ideological context, nature became an "enemy" that had to submit to human commands. Relying on ever-increasing material and intellectual resources, the modern states gradually gained control over nature, tempering the "wildness" of its unleashed forces.[18]

The works in the region of the Iron Gates have been examined within a similar theoretical vision. Luminita Gatejel has analyzed in detail the first technical project to make the gorge navigable, with a special interest in the way in which the hydraulic works resulted in a complex process of circulation of knowledge among experts in various corners of Europe and in how this knowledge was adapted to the particularities of the Danube project. To inform themselves about the state of knowledge in the field, Széchenyi and

---

16 Ibid., 163–180.
17 Mark Cioc, *The Rhine: An Eco-Biography, 1815–2000* (Seattle, 2002).
18 David Blackbourn, *The Conquest of Nature: Water, Landscape, and the Making of Modern Germany* (New York, 2007).

Vásárhelyi visited hydraulic works on the Rhine and then traveled in Bavaria, France, and Britain to consult with respected engineers and to purchase modern hydraulic equipment. Part of a continuous learning process, their plans were revised and adapted as the observation of already executed engineering works enabled a better understanding of the processes of nature.[19]

The Hungarian philologist Edit Király has also studied the technical works in the Danube Gorge in terms of the Foucauldian concept of "regularization." The regularization of the river was not a mere discursive subjection of a natural element considered unsafe but an integral part of the logic of civilization, of "correcting" nature. The conquest of nature was a "civilizing mission," an aspect all the more interesting today, when it is examined from the perspective of environmental history or the history of science and technology.[20] In the case of a great river that functioned as a natural highway for international trade, the organicist vision characteristic of the period considered that the natural obstacles (defiles, rocks, sandbanks) that impeded navigation were maladies that had to be operated on to enable healing. For navigation on the Danube, the Iron Gates constituted a serious dysfunction, the elimination of which would restore the health of the river organism. The attitude to nature was thus an ambivalent one: on the one hand, it was worthy of imitation, while on the other, it had to be controlled and improved.[21] Steam navigation was the technology that could reestablish economic circulation through a system that was currently sick, and the engineers were the surgeons who had to carry out the necessary interventions in the Danube Gorge. In her interesting analysis, Edit Király also puts an emphasis on the "dispositif" that permitted the works to be carried out—in other words, the vast political, economic, and institutional mechanism and the structures of knowledge engaged in the project of planning and implementing the hydraulic works.[22] The regularization of navigation was thus the result of complex technologies of power, with the natural environment being transformed into a major field of biopolitical intervention. The hydraulic works carried out in a region that had not previously known such engineering interventions constituted one of the most complex infrastructure projects of the nineteenth century in southeastern Europe.

---

19 Gatejel, "Overcoming the Iron Gates," 164–165.
20 Király, *"Die Donau ist die Form,"* 23–99.
21 Ibid., 96–97.
22 Ibid., 35.

Edit Király's book is a valuable source, with relevant examples, for an understanding of the functioning of such mechanisms of power, in which engineers became visible public actors because their work was more and more socially relevant. And the technical work in the Danube Gorge benefited from the mind of an extremely capable engineer.

ENGINEERING AND DIPLOMACY

When Széchenyi undertook to involve himself in the regularization of the Danube Gorge, he chose to work with a civil engineer, not a military one. Being a militarily and geopolitically sensitive zone, situated on the border of the Habsburg Empire, the region had hitherto been the exclusive province of military engineers. When he arrived with his civil collaborator, Széchenyi aroused considerable displeasure among military engineers, who were familiar with major infrastructure projects.[23] In addition to the other arguments in favor of his choice, the count's preference arose from a new vision of the future of the engineering profession, perhaps resulting from his knowledge of the Western world.

The engineer with whom Széchenyi had an excellent collaboration was Pál Vásárhelyi (1795–1846). After studying philosophy at Prešov, Vásárhelyi turned toward the more practical profession of engineering, graduating from the Faculty of Engineering of the University of Pest in 1816. In the years that followed, he specialized in hydraulic studies and works, published a number of articles on new methods of triangulation, and made hydrographic maps of the rivers Tisza/Tisa and Körös/Criş. In 1829, he was appointed to study the conditions of navigation on the Danube, and in 1833 he joined Széchenyi as a civil engineer for the regularization works on the Danube Gorge. He later became one of the most respected engineers in Hungary, a member of the Hungarian Academy and the author of daring plans to regularize the course of the Tisza.[24]

The carrying out of the hydraulic works also called for considerable diplomatic activity. Over and above the negotiations the cabinet in Vienna pursued through official channels in Istanbul, Belgrade, and Bucharest, Széchenyi

---

23   Gatejel, "Overcoming the Iron Gates," 164–165.
24   Anon., "Vásárhelyi Pál," https://dokutar.omikk.bme.hu/archivum/angol/htm/vasarhelyi_htm (accessed August 18, 2023).

carried on his own personal diplomatic activity, through which he persuaded decision-makers to support his project or at least not to obstruct it at the practical level. He had discussions with the Russian occupation forces in Wallachia, especially with General Pavel Kiselyov, and later with the ruler of the Principality, Alexandru Dimitrie Ghica. He also convinced the Serbian prince Miloš Obrenović (1780/3–1860, ruler of Serbia 1815–1839 and 1858–1860), who became a DDSG shareholder,[25] and, above all, he paid court to Hussein, the powerful pasha of Vidin. Ottoman agreement was decisive for the success of the initiative. It is interesting to note the different perspectives of the two empires regarding technical works: the Austrians were eager to enhance the power of their monarchy by the "taming" of the Danube, a work of great economic value, but also a symbol of civilization, while the Ottomans were opposed on military and religious grounds to intervention on the river bed, which meant disturbing the state of nature as it had been left by God. Széchenyi proved to be an excellent negotiator, and it was largely thanks to him that the DDSG steamboat services were able to operate in territories outside the Austrian monarchy.[26]

After a long European journey, Széchenyi and Vásárhelyi returned to the Danube Gorge in June 1834 and began the action of clearing the navigable channel with explosions in a number of rocky areas. As it was a dry year and the water level was low, Vásárhelyi managed to map the gorge, enabling a better understanding not just of the geography of the region but also of the complexity of the engineering works of regularization. As hydraulic works on the right bank were still difficult for diplomatic reasons, he began the construction of a road along the left bank, between Moldova Nouă and Orşova. There is no need to emphasize that, from a technical point of view, the works were complex and costly. Despite the cutting-edge technology, what was achieved in 1834 was still insufficient to make the region fully navigable, especially as the elimination of rocks had modified the speed of the river. Vásárhelyi had to update his technical project, which included the creation of a canal with two locks along the left bank.[27]

25 Gordana Karović, "Establishing Steam Navigation in the Principality of Serbia," in Dragana Amedoski (ed.), *Belgrade 1521–1867* (Belgrade, 2018), 385.

26 Miroslav Šedivý, "From Hostility to Cooperation? Austria, Russia and the Danubian Principalities 1829–40," *Slavonic and East European Review* 89, no. 4 (2011), 630–661.

27 Gatejel, "Overcoming the Iron Gates," 173–174.

In 1834, the reduced level of the Danube, by making obstacles on the river bed more easily visible, enabled the steamer *Argo* to continue downstream of the Iron Gates. This was a symbolic passage, which managed to repair the geographical fracture that had prevented the river from serving as the spine of Central and southeast European commerce.[28] The "new argonaut" had tamed the fury of nature; "the river and the Sea were opened to navigation"; Austria, Hungary, and Germany were unchained; "and before these lands with great productivity, flourishing possibilities were opening up."[29] The gorge was not completely passable, and as can be seen from Quin's account, the transfer of travelers and goods was complicated and often uncomfortable. However, from a symbolic point of view, the successful extension of the DDSG route to the Lower Danube was an important step toward the achievement of a connection between Vienna and Istanbul.

Vásárhelyi's works on the bed of the river were provisional and did not yet permit the safe circulation of the Austrian steamboats. Passage through the gorge depended on the depth of the river, and this varied according to the rhythms of nature. In August 1837, a steamboat managed to make the passage upstream too. As for the riverside road, this was completed in 1837,[30] enabling the connection to operate even in periods when the depth of the water was insufficient for passage by boat. In the following decades, naval engineers succeeded in building steamboats more suited to the conditions in the gorge. As mentioned in Chapter 1, vessels with lower draught began to circulate in the region, and already in the 1850s, the difficulties were much reduced. The Danube had been at least partially "tamed."

For economic reasons, however, much more was needed. Plans for regularization came to the fore again in the context of the Crimean War, and then in the 1870s. In 1879, the Hungarian government entrusted the engineers Ernö Wallandt (1845–1912) and Alajos Hoszpótzky (1857–1917) with the elaboration of the technical project, and the works were inaugurated with much pomp in 1896.[31] The last phase in the hydraulic transformation of the region was the construction of the two hydroelectric power stations—Iron

---

28 Anon., "Steam Navigation of the Danube," *Morning Chronicle* (December 10, 1834).
29 Thibault Lefebvre, *Études diplomatiques et économiques sur la Valachie* (Paris, 1858), 336–338.
30 Gatejel, "Overcoming the Iron Gates," 174–175.
31 Király, *"Die Donau ist die Form,"* 71–72.

Gates I and II—a gigantic infrastructure project of the communist period, which not only tamed nature but also partially drowned it.

"We Had Drifted from Wonders to Wonders"

The prolific French novelist Gabrielle Anne Cisterne de Courtiras, vicomtesse of Saint-Mars (1804–1872), a collaborator of Alexandre Dumas, is better known by her pseudonym "la Comtesse Dash." Among the dozens of books she published, several have a connection with the Romanian space. In her historical novel *Mikaël le Moldave*, translated into Romanian by Teodor Codrescu as *Mihail Cantemir Moldoveanul* (Iași, 1851), she describes the love between a Moldavian prince and a young Frenchwoman, whom the young Mihail Cantemir had met during his studies in the Paris of the Sun King. Her source of inspiration would appear to be her own relationship with Grigore Sturdza (1821–1901), nicknamed "Beizadea Vițel" ("Prince Calf"), the son of Mihail Sturdza, ruler of Moldavia from 1834 to 1849. While studying in Paris, Grigore Sturdza fell in love with the elegant countess, whom he brought to his homeland to become his wife. The lovers, Radu Rosetti tells us, settled at Perieni, "an estate in the vicinity of Iași, belonging to the *beizadea*'s mother. When the hospodar was informed by his son regarding his matrimonial intentions, he met them with a veto of the most categorical." The father's authoritarian intervention ruined the couple's clandestine marriage. The countess was taken to Galați, where, in keeping with the content of this book, "she found her place on a steamboat paid as far as Constantinople, together with a gift of money permitting her to continue in a fitting manner her journey as far as France."[32]

In 1866, the countess published a volume entitled *Les vacances d'une Parisienne*, in which she relates her voyage along the Lower Danube and then through Wallachia and the Balkans. The presentation is a collage of fact and fiction, imprecise as regards geography and time, but the details she gives about her voyage on the Danube suggest that it can safely be placed in the mid-1840s.

The description of the Danube Gorge occupies a special place in the countess's narration, in which she uses all her writerly abilities to capture

---

32 Rosetti, *Amintiri*, 179.

the beauty of a place "where the poetry of the soul awakens." She spent a night in Drencova, where the Danube seemed to have swallowed everything around it, and a thick forest invited meditation. These were special moments, as the weather was fine and "this wild nature lit by the moon had a special charm." The following day, an "imposing and grandiose" spectacle awaited the travelers:

> We were going to pass through what is called the Iron Gates, although there are neither iron nor gates. There are however rocks that hem in the river as far as the eye can see, it is magnificent. You are surrounded by little waterfalls, the boat floats on a bed of stones, reefs enclosed us on all sides, and the birds of prey wheeling around above our heads filled the air with their cries.
> This passage offers great dangers. It takes the habit and experience of the pilot to get through it. The uproar adds to the horror, or rather the beauty of the scene, you are so busy hearing, seeing, that you do not have time to be afraid. This pass and those of the Kazan are famous among sailors and among lovers of nature, it is surely one of the most beautiful spectacles that it is given to man to contemplate.[33]

Nature and history paraded before the travelers' eyes in a total experience. Close to Sviniţa, on the Serbian bank, the rock of Greben, protruding into the river, forced the Danube to make a tight bend. It was "something immense, majestic; one cannot admire these marvels without thinking of God who created them for us and without murmuring a prayer." A little further downstream could be seen the ruins of the castle of Tricule, erected in the fifteenth century by a local feudal baron to stop Turkish expansion (Figure 23). The travelers considered that its towers, "built irregularly, two on the summit of the rocks, a third down toward the river," had been built by the Romans as part of the infrastructure constructed by Trajan to conquer Dacia. "The Danube ceased to be hemmed in at this place, on the contrary it comes to be of splendid grandeur and majesty. The hills and the rocks were getting further away, although at the same time we could make out the famous caves that often gave refuge to warriors" in both ancient and

---

33   Comtesse Dash, *Les vacances d'une Parisienne* (Paris, 1890), 211–214.

Figure 23 Bartlett, *Drey-kule, or Tricola, Swinitza, with Roman Remains* (*c.*1840).

modern times. On the banks, there were mountains and forests, and "in the middle of the river, there rises, after one or two quite picturesque villages, an immense rock, in the middle of the river, which breaks against this obstacle, churning in foamy waves, and though the passage is easier, one still feels the murmuring breakers."[34]

The vessel then entered

> the admirable pass of Kazan—it is the name of the rock just mentioned— where one finds again the same wild setting as at the first Iron Gates. My God! How beautiful it is, there are no expressions to render it. Nothing can give an idea of it, the imagination would not know how to go so far, one has to have seen it.
>
> The mountains rise to a prodigious height, and they are rocks cut sheerly. The waves break at their feet with the same noise that was to accompany us during the whole journey. We were approaching a place where one speaks only with veneration, so to speak, of the Tables of Trajan [Figure 24].

---

34 Ibid., 214–215.

Figure 24 Bartlett, *Inscription on the Via Trajana* (*c.*1840).

Alas! Time has considerably mutilated them, but for all that, they can still be seen.[35]

The travelers admired the granite panel "carved to perpetuate the completion of the road and the victories of Trajan over the Dacians." Here, "in these lost lands," the monument produced a heightened effect as a link with the civilization of antiquity, and indeed with civilization in general. Trajan had chosen the splendid setting well, the countess thought. At Orșova, "the extreme frontier of Hungary," the picturesque took the form of human diversity, with the costumes of the various ethnic groups producing "a charming effect." The "half ruined" fortress on the island of Ada Kaleh also "produced a very picturesque effect in the landscape." Then

after the rapids, came the second Iron Gates, which could better deserve the name than the first, because of the ferruginous rocks by which the river is surrounded. The wild aspect became even more wild. Since our departure from Drencova, we had drifted from wonders to wonders, I do

35   Ibid., 215–216.

not believe that any river in Europe can offer places more beautiful and more unexpected.[36]

RITES OF PASSAGE

The Comtesse Dash's account of her journey is illustrative of the way in which the passage of the Danube Gorge using the service offered by the DDSG had become standardized and ritualized, with the passengers receiving a guided tour of the natural wonders, included in the price of their ticket. One of the advantages of regular steamboat services was the predictability of the journey, its reproducibility,[37] the fact that it was no longer a voyage subject to the caprices of the weather, but a route clearly demarcated and marked, with clear spatial and temporal stops. Each passenger knew what they were purchasing, and they could choose to repeat the experience of those who had gone before them. Travel guides compiled for their readers a list of picturesque and memorable places to admire, with the result that, in the passage of the Danube Gorge too, passengers could enjoy an experience at once unique and collective.

As a pioneer of travel in this sector of the Danube, Quin was an important point of reference in the writings of Western travelers, especially as the Murray guides included some extracts from his book. Murray was the principal source of documentation for anglophone passengers going through the gorge, offering precise information about what was interesting and picturesque to see. Already in the first edition (1837), the guide to the voyage through the Habsburg Empire provided a detailed account of the scientific, technical, and historical aspects of the route, interspersed with aesthetic appreciations of the region. Almost all the places that have been referred to here from Quin's and Comtesse Dash's books can be found in Murray too, along with details about the geological forces that formed the gorge, the rocks resembling a rhinoceros, local legends, caves, currents and torrents, and Golubac flies. There were also landscapes that failed to live up to expectations: "The rocky defile from Drenkova to the Greben [rock] is indeed grand; it was in it that Quin saw so many strange sights, which I could not re-discover." It

---

36   Ibid., 217–219.
37   Jonathan Stafford, "A Sea View: Perceptions of Maritime Space and Landscape in Accounts of Nineteenth-Century Colonial Steamship Travel," *Journal of Historical Geography* 55 (2017), 69–81.

was true, however, that the rocks had "a fantastic appearance, projecting forward like walls, or the side scenes in a theater, one behind another, sometimes rising upwards in the form of towers, battlements, and obelisks. Near the rapids the sailors pointed out one mass, which they call the 'Turk,' from some imaginary likeness." The place was spectacular and exceeded in grandeur any gorge through which the author had previously passed. The picturesque quality of the region, he considered, compensated for the monotony of the plains further upstream. History, too, contributed to the impression, and the traces of Roman civilization made the gorge all the more impressive.[38] The information was updated in the following edition (1840), including an extract from the description by Frederick John Monson, who had visited the region in October 1839. The guide quoted Monson's comparison to describe the Iron Gates: "a broad belt of low bristled rocks, like a vast harrow with the spikes upwards, which tears the shallow stream into countless adverse eddies."[39] The American Clemuel Green Ricketts (1822–1885), who traveled through Europe and the Levant in 1841–1842 and arrived in the region in May 1842, borrowed from Quin and Monson, as quoted in Murray's guide.[40]

Being so popularized, the place became a *must see*, sometimes giving rise to expectations that ended in disappointment. "I found these falls," noted the Austrian Ida Laura Pfeiffer, "and indeed almost every thing we passed, far below the anticipations I had formed from reading descriptions, frequently of great poetic beauty."[41]

The spectacular character of the region derived from a combination of other factors, starting from the manner in which travelers sailed downstream. Floating down the mighty river in an open boat, passengers could enjoy a true adventure,[42] a total sensorial experience. It was a sort of extreme rafting, in which the imminence of danger made the place all the more special. The sight was assaulted by the play of light between the mountains, with rapidly alternating colors, contrasts, and reflections. A "sea monster" splashed water, and the travelers could feel for themselves the waves and eddies of the river. Hearing played its part, too. Travelers' ears were filled with the fury of

38   *A Handbook* (1837), 377–389.
39   Monson, *Journal*, 160.
40   Clemuel Green Ricketts, *Notes of Travel: In Europe, Egypt, and the Holy Land, Including a Visit to the City of Constantinople, in 1841 and 1842* (Philadelphia, 1844), 221.
41   Pfeiffer, *Visit*, 29–30.
42   Richard Phillips, "Adventure," in Forsdick et al. (eds.), *Keywords*, 4–6.

the waters, the bubbling of the Danube, its eddies, and the roar of the wind and the current. After a crazy ride between the rocks of the Iron Gates, during which any connection with the outside world seemed to be interrupted, Baroness Aloïse-Christine de Carlowitz (1797–1863) recorded the feelings that took hold of her. The boat was making its own way between jagged rocks like the teeth of a beached whale, between eddies, rocks, whirlpools. Then it continued floating

> rapidly along the edge of a host of narrow waterfalls, separated one from another by formless rocks, like imaginary pedestals for the petrified giants who stretched out their fantastic hands above the waterfalls. You would have said that, under this horrifying appearance, they had kept their power of speech, such as is spoken only in Hell. The unclear voices that howled, whistled, groaned around us merged into a terrifying noise so powerful that not even a twenty-gun salvo would have been heard. On everyone's face a dumb unease was imprinted; only the helmsman was afraid of nothing. After weighing with his eyes all the falls of water, he dashed with a triumphant air toward the fastest of them.[43]

The feeble boat, ably handled by semi-barbarous helmsmen, was part of the spectacle. After sailing on one of the most modern vessels available at the time, the passengers were now able to appreciate the knowledge and practical experience of often illiterate helmsmen, whose skill could master the chaos of nature. The members of the crew also acted as guides, pointing out places of interest and making the necessary stops so that the beauty of the landscape could be appreciated in the best conditions. Their tip was often well earned.

The success of the route had the effect of ritualizing the passage through the gorge. Almost all travelers mentioned the same sites of interest and described the landscape in relatively similar ways, more often than not outdoing one another in superlatives. The considerable variation in the width of the river, now squeezed between mountains, now spreading like a lake in the plains, the impressive rocks and their legends of evil-doers and knights, spectacular slopes

---

43   Aloïse-Christine de Carlowitz, "Voyage dans les Principautés danubiennes et aux embouchures de Danube," *Revue de Paris* 30 (September 15, 1856), 535–536.

and ancient ruins, the prison of Empress Helena, the gnats of Golumbacz (Golubac) and the slaying of the dragon by Saint George, the rapids, the rock of Greben, the castle of Tricule, the Kazan gorge, the remains of the Roman road, Veterani's cave, Trajan's Tablet, the picturesque appearance of the locals, Orşova and its quarantine station, the baths of Mehadia, New Orşova (Ada Kaleh), the Iron Gates, the "hill of Alliom" (Alion), and Trajan's bridge, all were landmarks of a journey so memorable that it features in most accounts of travel in the region, beautified by the choicest epithets and metaphors.[44]

Romanian travelers seem to have been equally ready to praise the picturesque qualities of the region. Dimitrie Bolintineanu notes that

> the deep, mournful, wild lowing of the waves breaking against the rocks lying in their way all across the river added to the wild majesty of these places that had captured our eyes with a thousand forms. [...] The waters are like those beautiful women who appear much more beautiful in fury than in their peace. Nowhere does the Danube present a more enchanting appearance than between Orşova and Cladova, because of the cataracts. For me, I confess that this wild and imposing beauty can rival the most renowned places in the world.[45]

Nicolae Filimon describes the zone between "mountains of granite of a wild and severe beauty, through which the Danube flows furiously and groaning like an irritated lion." Going on his first journey abroad, the writer does not hide his admiration:

> Those who have not had the occasion to know how powerful and majestic is nature in her works and how enchanting and grand is her architecture—traveling from Orşova to Drencova—will surely be left in ecstasy seeing on both banks of the Danube superb pyramids of granite that form a multitude of poetic groups. In some places, the rocks are of a terrifying height and, through their position leaning toward the course of the river, seem to threaten the travelers; for all that, they have stood in this position since that last transformation of the earthly

---

44 *A Handbook* (1840), 449–461.
45 Bolintineanu, *Călătorii pe Dunăre şi în Bulgaria*, 4, 14.

globe, to show mankind, those little insects full of passion, how great and miraculous are the works of nature![46]

A Complicated Transfer

The transfer of travelers through the region did not always run smoothly. Mellen Chamberlain (1795–1839) was, in more than one respect, a visionary. A lawyer by profession, he went into business and, in the 1830s, became a prosperous entrepreneur specializing in the sale of weighing machines produced by the firm Fairbanks. In 1838, together with his wife and daughter, Chamberlain set out on a journey through Europe, which turned into a business opportunity when, in Paris, he met the inventor Samuel Morse (1791–1872). Morse was promoting his telegraph, and Chamberlain decided to invest in the sale of the device in Asia, Africa, and Europe. He proceeded to make a number of journeys in the region, and it seems that he managed to present the telegraph in Greece, Egypt, and the Ottoman Empire. At the beginning of 1839, he tried to obtain an interview with Sultan Mahmud II. As something was not working well in the prototype he had with him, he decided to go to Vienna to have it repaired.[47] He embarked on an Austrian steamer, and on May 14, 1839, he was drowned in the turbulent waters of the Danube Gorge, taking with him the prototype, the preliminary agreements that had been concluded, and Morse's hopes of capitalizing rapidly on his invention.

From the quarantine in Orşova, the British consul in Bucharest, R. G. Colquhoun, wrote that ten people had drowned out of the twenty-five souls on board the vessel. In addition to Chamberlain, the deceased were Stainsberg, the Austrian consul in Thessaloniki; his dragoman and two servants; the Scotsman Allen Roberts from Glasgow; the American lieutenant Pirie; an unidentified German; and two members of the crew.[48] The British officer

---

46 Filimon, *Escursiuni*, 16.

47 Yakup Bektas, "The Sultan's Messenger: Cultural Constructions of Ottoman Telegraphy, 1847–1880," *Technology and Culture* 41, no. 4 (2000), 671; Jane Alper, "VHS Awards Grant to Study Peacham Native Mellen Chamberlain," *Peacham Patriot. Peacham Historical Association* 33, no. 1 (2017), 3, 6, www.peachamhistorical.org/wp-content/uploads/2017/08/Patriot-Spring-2017-final-1.pdf (accessed August 28, 2023).

48 TNA, FO 78/363, f. 57–61 (Orşova, May 25, 1839).

James John Best, who sailed through the gorge several weeks later, noted the context in which the accident had taken place:

> It appears that from some unaccountable cause, and whether occasioned by accident or neglect it is difficult to ascertain, the barge in which the passengers and baggage were being towed up the stream, was brought by the force of the current with its broadside against it in its most rapid part. The force of the towing exerted against the short mast to which the tow-rope is attached, of course tending to make the barge lean over considerably on the side against which the current was pressing, the force of the current was too strong for the barge to withstand, and she was capsized in an instant, and all persons who were on the deck or raised part being immersed in the current, were carried down the rapid, and never more seen. Those who had remained in the interior of the barge were saved without the slightest injury, excepting a good wetting, the force of the men and oxen towing it bringing the barge alongside the bank of the river in an instant after the accident had occurred.[49]

As a result of this incident, the company introduced changes in its procedures in the interest of greater safety. In the most perilous sections of the gorge, where there was a risk of such accidents, the road on the left bank was used.

Some explanation is also called for regarding how the transfer along the gorge was carried out. The operation of the route between Vienna and Istanbul presupposed the organization of intermediary stations at key points in the region—namely Drencova, Orşova, and Schela Cladovei—to enable the transfer of travelers and goods with maximum safety and efficiency. Drencova, the initial terminus of the steamer from Budapest, was an important location on the map of Danube travel. The small inn built by the DDSG was the first place where travelers could sleep in a "normal" bed after several days on the river. Between Drencova and Schela Cladovei, the Danube passed over six rocky reefs. The channel was narrow and difficult, often less than 50 cm in depth, and the speed of the river was compared with that of a millstream. The steamers could pass through this section only when the river was high; for most of the year, the channel was impassable for vessels

---

49 Best, *Excursions*, 315–316.

of normal draught. The distance between Drencova and Schela Cladovei was thus covered in two stages, with a stop in the pleasant port of Orşova, close to the Austrian–Wallachian border. The journey might be made by road, or travelers might be transferred to flat-bottomed barges, which could also carry their luggage. Some travelers also refer to the sailing cutter *Tünde*, whose cabin could accommodate around twenty-five people.[50]

On the journey upstream, from Orşova to Drencova, luggage was sent on DDSG boats, while passengers were transferred either by boat or in carriages along the new road, "worthy to rank among the grandest works of modern times."[51] After the fatal accident mentioned above, travelers began to make the transfer by road, in the comfortable vehicles offered by the DDSG. This stage in the journey took around ten hours, and the road aroused travelers' admiration for the spectacular manner in which it was cut into the rock. In the view of the prelate Jacques Mislin, it was "well traced, artfully constructed, and [...] one of the finest works of this sort."[52]

Orşova was a pleasant resting place along the way, where travelers would wait for several days for the arrival of the steamer that was to take them downstream from Schela Cladovei and for the transfer of luggage and goods between the two steamboats. It offered good accommodation, but, as in Bratislava and Budapest, this was at passengers' expense. Many tourists took advantage of their stay here to visit the region, especially the baths of Mehadia (Băile Herculane), some 20 km away.

The transfer along the approximately 15 km from Orşova to Schela Cladovei (the Iron Gates section) was made either by road, in a carriage, with the journey taking around three hours, or, more commonly, in a flat-bottomed barge. In the latter case, the passengers "embarked on a new craft, the *Saturnus*, which is only covered in overhead, and is open on all sides."[53] On the way upstream, passengers were taken to the quarantine station of Orşova in open boats drawn by people or animals, under the escort of quarantine guardians and inspectors. According to Octavian Blewitt, who traveled this way in 1839, the journey upstream took nine hours.[54]

---

50 *A Handbook* (1837), 378–379.
51 Peregrine, "The Danube," 2, 684.
52 Mislin, *Les Saints Lieux*, 48.
53 Pfeiffer, *Visit*, 29.
54 Peregrine, "The Danube," 1, 570–571.

Things were much better in the 1850s, when the DDSG had vessels more suited to these sections.[55] Depending on the depth of the channel, the steamboat from Budapest now sailed directly to Schela Cladovei or, when this was not possible, a steamer with lower draught was used. The upstream journey, on a special steamboat, took five hours from Orşova to Drencova and seven hours from Orşova to Moldova Nouă.[56] In 1853, express steamers were introduced, which could sail the whole length of the Danube Gorge without difficulty, thus completely uniting, for travelers in a hurry, the Middle and the Lower Danube.

Mehadia offered a relaxing excursion for travelers passing through the Danube Gorge, who could take advantage of the time needed for the transfer of their luggage to enjoy a destination that was still little known at the European level. Roman ruins had been discovered there in 1734 by General Johann Andreas von Hamilton (1679–1738), the commander of the imperial troops in the Banat. Gradually, a settlement developed for medicinal purposes, initially for the soldiers of the border regiment and later for civilian customers.[57] The thermal baths were recommended for a wider range of ailments, including—according to the British travel writer Edmund Spencer, who passed through the region in 1836—"chronic cases of scrofula, cutaneous diseases, rheumatism, gout, contractions of the limbs, consumption of the lungs, diseases of the eyes, &c." The "sanative properties" of the water were not the only attraction of the place, Spencer remarks, as "the surrounding country is beautiful, abounding with romantic valleys and lofty hills"[58] (Figure 25).

With the coming of steam navigation, Mehadia was now on a much-circulated route, and more and more tourists began to visit it. Those sailing on the Danube did not come so much to make use of its spa facilities as to enjoy a day on dry land in the midst of nature and history. Their descriptions popularized the place, turning it into a favorite holiday location for the elites of Hungary and the Danubian Principalities. Tourists generally observed

Figure 25 Bartlett, *Baths of Mehadia* (*c.*1840).

two things: the landscape and the people. "The situation is sequestered and delightful; falling waters, picturesque rocks, and groves enliven and embellish the spot," writes the British naval officer Adolphus Slade and goes on to mention some of the Roman vestiges to be seen.⁵⁹ The human factor was equally attractive, with the mixture of locals and tourists characteristic of a modern resort: ladies in the latest fashions, and peasant women in their traditional costume.⁶⁰

DANUBE VIEWS

"A picture is worth a thousand words" goes a saying that was famous long before Instagram was thought of. The nineteenth century was also an age of the visual, as it became increasingly simple and cheap to produce and consume illustrations. In the last decade of the previous century, Alois

---

59 Slade, *Travels, 171.*
60 de Valon, *Une année,* 235–236.

abounded in "cities, palaces, castles, convents, churches—the splendid memorials of feudal and monastic times—its battle-fields and romantic forests," all of which deserved to be made known to the public. The magnificent landscapes were enriched and animated with scenes and monuments recalling ancient times. The region was full of tales and traditions that enriched the landscape and awakened living memories in the mind of the traveler. The illustrations offered a selection of the most remarkable and picturesque scenes encountered along the great river: "They present in turn, the magnificence of nature, the splendour of art, with striking contrasts and combinations of landscape, relieved by those minor features of national distinctions, which invest them with the characteristic traits of originality."[69] The volume seems to be directed both at prospective travelers and at past tourists wishing to keep a visual record of the places they had visited.

The Danube Gorge occupies an important place in the economy of the work. A number of engravings present points of interest that have been mentioned in this chapter: the Iron Gates, the Babacai rock, the ruined castle of Golubac, the fortress of Tricule and the Roman ruins, the entry to the Kazan pass, the Roman roads, Trajan's Tablet, a wedding at Orşova, the baths at Mehadia, the village of Kladovo, the remains of Trajan's bridge. Then there are small wood-engraved sketches of the steamer station at Drencova, the baths at Mehadia, and the channel in the gorge toward Mehadia.

In all these illustrations and in the accompanying text, we may notice the mixture of the picturesque and storytelling to attract the reader. The presentation of the Babacai rock is one example, and indeed its story is told by many other travelers (Figure 26). The rock rose abruptly, in splendid isolation in the middle of the river. Its name was derived from a dispute at once domestic and political. A Turkish ağa learned that Zuleika, the most beautiful of his wives, had run off with a Hungarian nobleman. Furious at the double treachery, the ağa instructed a janissary to bring back Zuleika and the head of her lover. Disguising themselves as Serbian peasants coming to complain about Turkish abuses, the janissary and his companions captured Zuleika and beheaded the Hungarian. Zuleika was left on a rock in the middle of the river, with the words "Ba-ba-kaÿ"—"Repent of thy sin!"—resounding in her ears. However, it turned out that in the heat of the moment, the

---

69   Ibid., 1–2.

Figure 26 Bartlett, *Babacai* (*c.*1840).

janissaries had beheaded not the count but one of his aides-de-camp. The noble lover immediately set out to free his beloved. As another janissary, sent to end Zuleika's repentance and her life, could not find her on the lonely rock, the Turks assumed that she had drowned in the Danube. It was after himself being taken prisoner that the ağa learned the truth that came as a greater blow to him than all those he had received in battle: Zuleika had escaped, renounced Islam, and become her liberator's wife.[70]

Equally interesting was the castle of Golubac, another place that combined wild nature and Romantic landscape. The ruins of the castle illustrated the tumultuous history of the region. According to legend, it was there that the Empress Helena had been imprisoned—probably Helena Dragaš (1372–1450), wife of the Byzantine emperor Manuel II Palaeologus and mother of the last Byzantine emperors. Of particular interest was the cave "in which, according to tradition, St. George slew the Dragon—and whose carcass, it is said, still putrefies in its recesses, and sends forth those myriads of small flies, which are so tormenting to men and cattle." The malefic power of the beast could

70 Ibid., 204–206.

Figure 27 Bartlett, *Sozorney, with Remains of Trajan's Bridge* (c.1840).

still be felt in the seasonal raids of this "Kolumbacz-fly," an insect whose attacks were the stuff of grim stories.[71]

Everywhere in the gorge, the landscape displayed the "wild, solitary grandeur" of nature. From a boat steered by able helmsmen, passengers could admire more and more picturesque sites. The entry to the Kazan was spectacular, with sky, water, and rock "presenting, in the most striking combination, all those qualities, features, and appearances which are the essential constituents of sublimity in natural landscape." Interesting, too, were the testimonies in stone to "Roman skill and perseverance," Trajan's road and bridge (Figure 27). The volume continues with an account of Orşova and its ethnic diversity, with the description of a wedding there shifting the theme of the picturesque and the diverse toward the human communities of the region.[72]

Once the mountains were left behind, the riverside landscape was radically transformed, signaling the fact that a symbolic frontier had been crossed. The passengers now entered "one of those vast monotonous plains

71   Ibid., 206–207.
72   Ibid., 207–213, 221–222.

through which the Danube, split into numerous channels, and enclosing many islands, pours its capricious flood towards the Black Sea." It seemed the setting of another world, "presenting to the mind and eye a picture of unbounded solitude and desolation."[73] This did not, however, mean that the land downstream was without artistic interest, as was demonstrated by the illustrations of Nikopol, Ruse, and Sulina.

Conclusions

The example of the Danube Gorge serves to underline the direct connection between modern means of transport, the commodification of travel, and the perception of the natural landscape. The "discovery" of a picturesque region, generally presented by travelers in superlative terms, came almost simultaneously with the beginning of the transformation of the river into a circulated route of international mobility. The establishment of the steamboat service between Vienna and Istanbul favored the process of the aesthetic capitalization of a fascinating space, which had been begun by the publisher Kunike. Thanks to the hydraulic works and logistic arrangements coordinated by Vásárhelyi and Széchenyi, passengers were more able to appreciate the beauty of nature, a component that was included in the travel offer of the DDSG, thus marking a new form in which the natural environment was tamed and commodified.

The Danube Gorge contains numerous memorable sites: bridges and caves, fortresses and rocks, ruins and mountains. Seen as the opposite of civilization, nature is described as having a distinctive beauty precisely by virtue of its having remained wild, outside human control. However, it was the very intervention of civilization that made possible the revelation of this special space, which called for a thorough admiration with that "gaze" specific to the tourist. The way in which the region was crossed made its appreciation a total, multisensorial experience. The delight with which travelers experienced the passage through the gorge was part of a form of ritualization of the consumption of the beautiful, shifting admiration from the private to the public space. Passing through the region by a DDSG steamer had become a veritable public ceremony, in which the beautiful was shared with one's traveling companions.

---

73 Ibid., 221–222.

The passage was also emblematic because of the spatial and temporal significance of the region. A number of elements, including the application of the terms "gate" and "threshold" with reference to the Danube, heightened the sense that passengers were passing between worlds. Apart from the fact that it formed the border between neighboring empires, the gorge marked a symbolic passage between the West and the East and also, in terms of the river's flow, between upper and lower. The passage was at the same time a temporal one, with many vestiges of the Romans' material civilization to be seen, together with the attempts of contemporaries to copy their technical mastery, all contributing to the sense of an area of transition, a special zone, or, as one Italian traveler put it, a space that was "magical, miraculous."[74]

The admirable novel *Az arany ember* ("The Golden Man") by Mór Jókai (1825´–1904) captures best of all this special identity of the area, in which the river not only kept its grandeur and wildness but also constructed its own physical geography:

> Here, it has carved islands out of the stubborn granite, new creations, to be found on no chart, overgrown with wild bushes. They belong to no State—neither Hungary, Turkey, nor Servia; they are ownerless, nameless, subject to no tribute, outside the world. And there again it has carried away an island, with all its shrubs, trees, huts, and wiped it from the map.[75]

74 Giorgio Smancini, *Scorsa piacevole in Grecia, Egitto, Turchia sul Danubio e da Vienna alla Lombardia* (Milano, 1844), 141.
75 Maurus Jokai [Mór Jókai], *Timar's Two Worlds* (*Az arany ember*), trans. [Agnes] Hegan Kennard (New York, 1895), 3.

Conclusions

Various theories of modernization maintain that progress is not linear but rather takes place in leaps, with breaks in its rhythm and significant variations depending on space and historical time. The use of mechanical propulsion in the transport of people and goods was one such disruptive factors; it accelerated modernization wherever railways were constructed, and there were ports served by steam navigation companies. The Romanian case is illustrative of the way in which connection to the world and to modernity was produced by such means. The introduction of steam navigation on the Danube, a river little used as a route of international mobility before 1830, profoundly transformed the geopolitical situation of Wallachia and Moldavia. Now known as the "Danubian Principalities," the two states bound their destiny even more closely to navigation on the river, a transport artery that gave them both political importance and economic prosperity.

This book has examined in detail these changes, which may be summed up in terms of three dimensions with reference to (1) the transformation of the Danube into an international transport infrastructure and a connection along the West–East axis, (2) the transformation of the steamboat into a global space of intense socializing, and (3) the transformation of space and time in the Danube region as a consequence of the establishment of the regular steamer service between Vienna and Istanbul.

The transformation of the Danube into a navigable river along more than 2,000 km of its length and its use as a connecting route between Central Europe and the Orient was the cumulative result of a series of political, economic, and technological factors. Similar projects for the economic unification of the Danube had existed in previous centuries too; after 1830, a new international context and revolutionary technological inventions were creating new opportunities for the Austrian Empire to use the valley of the Danube as a direction of economic and political expansion toward the region of the Danube mouths and, via the Black Sea, toward the Ottoman Empire. Major geographical mutations were taking place in the region of the Black Sea, open to international trade thanks to the anti-Ottoman campaigns of the Russian Empire, which had annexed the territories on its norther shores and was threatening the Ottoman provinces in the Caucasus and the Principalities of Wallachia and Moldavia. Russia gradually transformed the economic value of the Black Sea: the string of ports founded or revived along its shores and those of the Sea of Azov created excellent business opportunities for merchants from all over the world. The growing Russian pressure on Istanbul, which lay only a few days' distance from the Crimean ports, brought with it growing threats to the strategic balance in the region of the Turkish straits. The Peace of Adrianople (1829), after the defeat of the Sublime Porte in a new Russian–Ottoman conflict, further consolidated Russian power, while the Convention of Hünkâr İskelesi (1833)—under which Tsar Nicholas I provided support to Sultan Mahmud II, who was threatened by his own vassal, Mehmet Ali, pasha of Egypt—turned the Eastern Question into the most serious threat to the balance of power on the continent.

With the eyes of European diplomacy focused on Istanbul, the Viennese authorities looked very favorably on the intentions of a private company to develop steam navigation on the Danube. Founded in Vienna, the DDSG was supported directly and indirectly by the government to extend its services beyond the Habsburg territories and to establish a connection between Vienna and Istanbul. The achievement of this connection in 1836, just six years after the steamboat *Franz I* had made its first voyage between Vienna and Budapest, was a resounding success and underlined the benefits of public–private partnership in such strategic projects.

The DDSG had invested considerable sums in importing the latest technology from the West, in building steamboats in the Erdberg shipyard, and in organizing a complex network of agencies to enable the service to operate in the intermediate ports and especially in the Levant.

The Austrian state made a substantial contribution in two other aspects: the transformation of the Danube into a transport infrastructure, and the reforming of the Austrian quarantine system, a compromise solution to the conflict between strict antiepidemic prevention measures and the ever-growing mobility of people and goods. With regard to the former, the authorities dealt with the navigability of the river in several critical sectors, the most important area of intervention being the Iron Gates Gorge. Although they did not succeed in fully "taming" the region, the hydraulic works coordinated by Count István Széchenyi and the engineer Pál Vásárhelyi allowed the DDSG to connect two hitherto fractured sectors of the river and thus made possible the extension of the steamboat service to the Lower Danube and into the Black Sea. As for the quarantine system, this was made more efficient both through collaboration with the Ottoman Empire and the Danubian Principalities and through a more efficient management of information. Toward the middle of the nineteenth century, lazarettos became much less of an obstacle in the way of the mobility of people and goods, due not only to the pressure of powerful shipping companies like the DDSG and Austrian Lloyd but also to their contribution to the construction of transport networks by means of which information circulated, rapidly and predictably, from the regions most directly exposed to health threats.

The Danube route constituted a transport infrastructure that permitted safe and fast travel between the West and the East. As travel by land through the Balkans was still dangerous, costly, and uncomfortable, the voyage between Vienna and Istanbul, part of an industrialized and commodified transport service, offered a combination of important advantages: speed, comfort, financial accessibility, and also safety. The mechanization of traction and the relatively decent level of comfort on board meant that there was no more need for a great consumption of human energy, and the travelers' bodies were no longer subject to the physical and mental fatigue caused by the constant shaking of a journey by road. Predictability of timing and cost was part of this form of comfort and of a new management of time that enabled travelers to make clear plans of their journey, starting from a timetable that was generally respected and costs that were always known in advance. The price of tickets was accessible, opening up the possibility of mobility to members of all social categories. Safety was also included in the price of the ticket, an important aspect in view of the fears aroused by the dangers to be encountered on the roads of the Ottoman Empire.

The steamer itself is the second important element of this equation. As a global phenomenon, travel means more than modern infrastructure, fast vehicles, and powerful means of propulsion: it comes also with an inherent social dimension that evolved concomitantly with the modernization of public transport. The transport revolution of the nineteenth century not only brought greater speed, increased comfort, and prices to suit all pockets but also determined new forms of social meetings for the passengers who embarked on a steamer or in a railway carriage. In line with new directions of analysis in the history of transport, the steamboats have been included in the model of mobile sociability. Sliding between various different worlds, they served as contact zones, functioning as a connecting space between two of the world's most attractive cities. According to various contemporary descriptions, the cabins and decks of the Danube steamers were veritable micro-planets where global actors gathered as a consequence of their need to travel. For the majority of writers, the cabin was the center of this universe, where the financially more powerful passengers met, interacted, and frequently formed a mobile community for which sociality was an integral part of the travel experience. Life in this physical space favored not only the exchange of precious information about the various stages of the journey but also the sharing of diverse social practices and norms.

In addition to its characteristics as a means of water transport, the steamboat is a social technology, a mobile salon circulating through space and time. As an effervescent medium of global connection and sociality, it is a perfect example of the way in which mobility and sociality combine and transcend the already flexible boundary between the private and public spaces. Detained on board for a period of at least several days, the travelers explored new "mobile temporalities," fixed in the ad-hoc practices of the cosmopolitan community that varied from the hour of dinner to that of sleep and which pushed passengers to adapt to the conviviality of their means of transport. Seen through the prism of such personal and subjective descriptions, the steamboat is a heterotopia, a fascinating theater of global history, a connector and a spatial, cultural, and social mediator of the world. The situation of Moldavian and Wallachian passengers is all the more interesting, as the steamboat was one of the most visible places in which Eastern European elites learned to be part of the "civilized world."

A third aspect concerns the consequences of the establishment of the steamboat service on space and time in the Danube region. Eliminating a number of major sources of physical and mental stress for passengers, the

Danube voyage became an interesting social experience and an aesthetic spectacle in which passengers could enjoy the cosmopolitan company they found themselves in and in which they had the leisure to admire the landscape unfolding around them. Looking along the banks of the Lower Danube, travelers discovered an interesting territory, which they "knew" both from their summary direct observation and, above all, from travel guidebooks. That the regional realities were "produced" by the steamboat emerges from a number of examples in which reference has been made to the clichés and stereotypes used to speak about the image of the Danube ports or the agricultural situation of the Principalities. A more interesting case is that of the representation of the Danube Gorge, where travelers show an interest in nature, ruins, history, and Roman civilization characteristic of the romantic period, constructing an image founded on certain elements considered "picturesque" and worthy of admiration. Time, in its turn, was compressed and reinvented by the steamboat and on the steamboat. Partly because the vessel also moved through time, allowing the passengers to feel the different historical periods in which the different regions through which the river passed seemed to belong.

The route drew the Danubian Principalities into the networks of international mobility. More and more travelers came to pass through the Principalities (or rather along their borders), while numerous Moldavians or Wallachians traveled to the West and the East on board the Austrian steamboats. What is remarkable is the rapidity with which this transport network was constructed and how easily the Romanians seem to have become used to traveling. The Danube ports, already part of a commercial revolution, profited fully from their status as hubs of one of the most dynamic routes of pan-European transport.

The route was also important for the modernity of commodified service and for the way in which it became part of the race to compress space and time. The circling of the world in eighty days began with the crossing of Europe from the West to the East in just two weeks, with the interlinking of various routes, nodes, and connections. The Principalities became connected to the rest of the world through the Danube route, which remained crucial to the Romanian economy for the next half century. It was only toward the end of the century that their dependence on the Danube route was reduced, after the construction of railways and the integration of Dobrogea in the

Romanian state. 1895 was symbolic as the year in which—after the construction of the Cernavodă bridge across the Danube and the equipping of a national shipping company (the Romanian Maritime Service) with modern vessels—the establishment of a rail connection through Constanța with transfer to the Black Sea marked the completion of the fastest connection available between West and East. The Danube was completely disconnected from this route, which was primarily based on rail transport. The great river's day had passed.

Bibliography

Unpublished Sources

The National Archives of the United Kingdom, Foreign Office, files from the funds 78 and 195.

Published Sources

Acharya, Malasree Neepa. "Cosmopolitanism." In Noel B. Salazar and Kiran Jayaram (eds.), *Keywords of Mobility. Critical Engagements*. New York: Berghahn, 2016, 33–54.

Ackerknecht, Erwin H. "Anticontagionism between 1821 and 1867: The Fielding H. Garrison Lecture." *International Journal of Epidemiology* 38, no. 1 (2009): 7–21. https://doi.org/10.1093/ije/dyn254.

Ainsworth, Francis W. "The Communication between the Danube and the Black Sea." *Mirror of Literature, Amusement and Instruction* 2 (September 1842): 152–154.

Ainsworth, Francis W. "Herr Andersen." *Literary Gazette* 1551 (October 10, 1846): 877–878.

Alecsandri, Vasile. "Înecarea vaporului Seceni pe Dunărea." In Alexandru Marcu (ed.), *Proză. Povestiri. Amintiri romantice*. Craiova: Ed. Scrisul Românesc, 1939, 73–85.

Alecsandri, Vasile. *Opere complete*, vol. 2: *Teatru I*. Bucuresci: Institutul de Arte Grafice şi Editură "Minerva," 1903.

Alexander, James Edward. *Travels from India to England ... etc. in the Years 1825–26*. London: Parbury, Allen and Co., 1827.

Allen, Esther. "'Money and Little Red Books': Romanticism, Tourism, and the Rise of the Guidebook." *Lit: Literature Interpretation Theory* 7, nos. 2–3 (1996): 213–226.

Alper, Jane. "VHS Awards Grant to Study Peacham Native Mellen Chamberlain." *Peacham Patriot: Peacham Historical Association* 33, no. 1 (2017): 3, 6. www.peachamhistorical.org/wp-content/uploads/2017/08/Patriot-Spring-2017-final-1.pdf (accessed August 28, 2023).

Amăriuţei, Mihai-Cristian, and Benone Dorneanu. "Spătarul Sandu Miclescu de la Şerbeşti (1804–1877). Schiţă de portret a unui 'paşoptist' uitat, pe baza unor documente şi amintiri de familie." *Archiva Moldaviae* 10 (2018): 69–90.

Analele Parlamentare ale României, vol. 16, part 1: *Divanul Obştesc al Ţărei Româneşti, Legislatura V, Sesiunea I (XV), 1850–1851*. Bucureşti: Imprimeria Statului, 1909.

Anastasiu, Florian, Vasile Anton, and Ştefan Cocioabă. *Monografia judeţului Brăila*. Brăila: Comitetul Judeţean P.C.R., 1971.

Andersen, Hans Christian. *A Poet's Bazaar*, vol. 3. Translated from the Danish by Charles Beckwith Lohmeyer. London: Richard Bentley, 1846.

Andersen, Hans Christian. *The Story of My Life*. Boston: Houghton Mifflin, 1871.

Angelo, Anastas. "Parakhodŭt 'Ferdinand I' po liniyata Konstantinopol–Varna–Sulina–Galats." http://morskivestnik.com/compass/news/2018/072018/images/ssFerdinand_08072018.pdf (accessed August 28, 2023).

Anghelescu, Mircea. *Lâna de aur. Călătorii şi călătoriile în literatura română*. Bucureşti: Cartea Românească, 2015.

Anim-Addo, Anyaa, William Hasty, and Kimberley Peters. "The Mobilities of Ships and Shipped Mobilities." *Mobilities* 9, no. 3 (2014): 337–349. https://doi.org/10.1080/1745010 1.2014.946773.

Anon. "Blyth's Novel Steamers on the Danube." *Daily News*, London, December 30, 1854.

Anon. "The Danube." *Fraser's Magazine for Town and Country* 49, no. 293 (May 1854): 575.

Anon. "The Danube Ship Canal, and a Free Port in the Black Sea." *Leeds Mercury*, Leeds, November 10, 17, and 24, 1855.

Anon. "Die erste k. k. pr. Donau-Dampfschiffahrt-Gesellschaft. Erste Periode. Gründung und Betriebs-Verhältnisse bis zum Jahre 1841." Centralblatt für Eisenbahn und Dampfschifffahrt in Österreich. *Wien*, 35 (August 30, 1862): 337–340;

Anon. "Die erste k. k. pr. Donau-Dampfschiffahrt-Gesellschaft. Zweite Periode. Vom Jahre 1842–1851." *Centralblatt für Eisenbahn und Dampfschifffahrt in Österreich. Wien*, 36 (September 6, 1862): 345–348; 37 (September 13, 1862): 353–355; and 40 (October 4, 1862): 377–379.

Anon. "The Different Roads from the Lower Danube to Constantinople." *Morning Chronicle*, London, July 30, 1828.

Anon. "Introduction of Steam Navigation into Austria." *Mechanics Magazine, Museum, Register, Journal and Gazette* 48 (1848): 103.

Anon. "The Mouths of the Danube, from the Notes of a Recent Traveller." *Colburn United Service Magazine. A Naval and Military Journal* no. 2 (1844): 189–202.

Anon. "Navigation of the Danube." *Army and Navy Chronicles* 2–3 (1836): 104.

Anon. "România după 20 de ani: Am fi putut hrăni toată Germania, dăm de mâncare la doar un sfert de Românie." *Capital*, December 21, 2009. www.capital.ro/romania-dupa-20-de-ani-am-fi-putut-hrani-toata-germania-dam-de-mancare-la-doar-un-sfert-de-r.html (accessed August 18, 2023).

Anon. "România poate să producă hrană pentru 80 de milioane de oameni, dar agricultura e de subzistenţă." *Newsweek România*, April 8, 2019. https://newsweek.ro/economie/romania-poate-sa-produca-hrana-pentru-80-de-milioane-de-oameni-dar-agricultura-e-de-subzistenta (accessed August 18, 2023).

Anon. "Steam Navigation of the Danube." *Morning Chronicle*, London, December 10, 1834.

Anon. "Steam Navigation of the Danube." *The Era*, London, January 23, 1853.

Anon. "Vásárhelyi Pál." https://dokutar.omikk.bme.hu/archivum/angol/htm/vasarhelyi_p.htm (accessed August 18, 2023).

Anul 1848 în Principatele Române: acte si documente, vols. 3 and 5. Bucuresci: Instit. de Arte Grafice "Carol Göbl," 1902 and 1904.

Ardeleanu, Constantin. *The European Commission of the Danube (1856–1948). An Experiment in International Administration*. Leiden: Brill, 2020. https://doi.org/10.1163/9789004425965.

Ardeleanu, Constantin. *Evoluția intereselor economice și politice britanice la gurile Dunării (1829–1914)*. Brăila: Editura Istros–Muzeul Brăilei, 2008.

Ardeleanu, Constantin. "From Vienna to Constantinople on Board the Vessels of the Austrian Danube Steam-Navigation Company (1834–1842)." *Historical Yearbook* 6 (2009): 187–202.

Ardeleanu, Constantin. *Gurile Dunării—o problemă europeană. Comerț și navigație la Dunărea de Jos în surse contemporane (1829–1853)*. Brăila: Editura Istros – Muzeul Brăilei, 2012.

Ardeleanu, Constantin. *International Trade and Diplomacy at the Lower Danube: The Sulina Question and the Economic Premises of the Crimean War (1829–1853)*. Brăila: Editura Istros–Muzeul Brăilei, 2014.

Ardeleanu, Constantin. "'Steamboat Sociality' along the Danube and the Black Sea (Mid-1830s–Mid-1850s)." *Journal of Transport History* 41, no. 2 (2020): 208–228. https://doi.org/10.1177/0022526620908258.

Armstrong, John, and David M. Williams. "The Steamboat and Popular Tourism." *Journal of Transport History* 26, no. 1 (2005): 61–77. https://doi.org/10.7227/TJTH.26.1.4.

Armstrong, John, and David M. Williams. "The Steamship as an Agent of Modernisation, 1812–1840." *International Journal of Maritime History* 19, no. 1 (2007): 145–160. https://doi.org/10.1177/084387140701900108.

Arnim, Karl Otto Ludwig von. *Flüchtige Bemerkungen eines flüchtigen Reisenden*, vol. 3. Berlin: Nicolaische Buchhandlung, 1837.

Arrizabalaga, Jon, and Juan Carlos García-Reyes. "Contagion Controversies on Cholera and Yellow Fever in Mid Nineteenth-Century Spain: The Case of Nicasio Landa." In John Chircop and Francisco Javier Martínez (eds.), *Mediterranean Quarantines, 1750–1914. Space, Identity and Power*. Manchester: Manchester University Press, 2018, 170–195.

Arslan, Aytuğ, and Hasan Ali Polat. "Travel from Europe to Istanbul in the 19th Century and the Quarantine of Çanakkale." *Journal of Transport & Health* 4 (2017): 10–17. https://doi.org/10.1016/j.jth.2017.01.003.

Audin, J.-M.-V. *Guide classique du voyageur en Europe*. Deuxième édition. Paris: L. Maison, 1852.

Bagwell, Philip. *The Transport Revolution, 1770–1985*. London: Routledge, 1988.

Bălcescu, Nicolae. *Scrisori către Ion Ghica*. Edited by Petre V. Haneș. București: Editura Librăriei Leon Alcalay, 1911.

Baldwin, Peter. *Contagion and the State in Europe, 1830–1930*. Cambridge: Cambridge University Press, 1999.

Bărbulescu, Constantin. *Physicians, Peasants and Modern Medicine*. Translated by Angela Jianu. Budapest: Central European University Press, 2018.

Barkley, Henry C. *Between the Danube and the Black Sea; or Five Years in Bulgaria*. London: John Murray, 1876.

Bashford, Alison (ed.). *Quarantine: Local and Global Histories*. Basingstoke: Palgrave Macmillan, 2016.

Beattie, William. *The Danube: Its History, Scenery, and Topography*, splendidly illustrated from sketches taken on the spot by Abresch and drawn by W. H. Bartlett. London: George Virtue, 1844.

Bejan, Cezar, Alexandru Duță, Stelian Iordache, and Viorica Solomon (eds.). *Tezaur documentar gălățean*. București: Direcția Generală a Arhivelor Statului, 1988.

Bektas, Yakup. "The Sultan's Messenger: Cultural Constructions of Ottoman Telegraphy, 1847–1880." *Technology and Culture* 41, no. 4 (2000): 669–696. https://doi.org/10.1353/tech.2000.0141.

Best, J. J. *Excursions in Albania*. London: Allen, 1842.

Bezio, Kelly. "The Nineteenth-Century Quarantine Narrative." *Literature and Medicine* 31, no. 1 (2013): 63–90. https://doi.org/10.1353/lm.2013.0007.

Bibliografia analitică a periodicelor românești, vol. 1: *1790–1850*, edited by Ioan Lupu, Nestor Camariano, and Ovidiu Papadima; vol. 2: *1851–1858*, edited by Ioan Lupu, Dan Berindei, Nestor Camariano, and Ovidiu Papadima. București: Editura Academiei Republicii Socialiste România, 1966–1972.

Bijker, Wiebe, and John Law (eds.). *Shaping Technology/Building Society: Studies in Sociotechnical Change*. Cambridge, MA: MIT Press, 1992.

Bijsterveld, Karin. "The Diabolical Symphony of the Mechanical Age: Technology and Symbolism of Sound in European and North American Noise Abatement Campaigns, 1900–40." *Social Studies of Science* 31, no. 1 (2001): 37–70.

Bissell, David. "Moving with Others: The Sociality of the Railway Journey." In Phillip Vannini (ed.), *The Cultures of Alternative Mobilities: Routes Less Travelled*. Farnham: Ashgate, 2009, 55–69.

Blackbourn, David. *The Conquest of Nature: Water, Landscape, and the Making of Modern Germany*. New York: Norton, 2007.

Bodea Cornelia, C. "Călătoria lui Bălcescu pe Dunăre în 1852." *Studii* 10, no. 1 (1957): 161–170.

Bohls, Elizabeth A. "Picturesque Travel: The Aesthetics and Politics of Landscape." In Carl Thomas (ed.), *The Routledge Companion to Travel Writing*. London: Routledge, 2015, 266–277.

Boia, Lucian. *Romania, Borderland of Europe*. Translated by James Christian Brown. London: Reaktion Books, 2001.

Bolintineanu, Dimitrie. *Călătorii*, vols. 1–2. Edited by Ion Roman. București: Editura pentru Literatură, 1968.

Booker, John. *Maritime Quarantine. The British Experience, c. 1650–1900*. London: Routledge, 2016.

Bossy, R. V. "Un drumeț danez în Principate." *Analele Academiei Române, Memoriile Secțiunii Istorice. Seria III* 24 (1941–1942): 1–8.

Botez, Constantin, Demetru Urma, and Ioan Saizu. *Epopeea feroviară românească*. București: Editura Sport-Turism, 1977.

Boué, Ami. *La Turquie d'Europe ou observation sur la géologie, l'histoire naturelle, la statistique, les mœurs, les costumes, l'archéologie, l'agriculture, l'industrie, le commerce, les gouvernements divers, le clergé, l'histoire et l'état politique de cet Empire*, vol. 3. Paris: Librairie de la Société de Géographie, 1840.

Boué, Ami. *Recueil d'itinéraires dans la Turquie d'Europe*, vol. 1. Vienne: W. Braumüller, 1854.

Bracewell, Wendy (ed.), *Orientations: An Anthology of East European Travel Writing: ca. 1550–2000*. Budapest: Central European University Press, 2009.

Bracewell, Wendy. "Travels through the Slav World." In Wendy Bracewell and Alex Drace-Francis (eds.), *Under Eastern Eyes. A Comparative Introduction to East European Travel Writing in Europe*. Budapest: Central European University Press, 2008, 147–194.

Braudel, Fernand. *Civilization and Capitalism, 15th–18th Century*, vol. 1: *The Structures of Everyday Life*. Translated by Siân Reynold. Berkeley: University of California Press, 1981.

Brother Peregrine (Octavian Blewitt). "The Danube." Part 1 and Part 2. *Fraser's Magazine for Town and Country* 22, no. 131 (November 1840): 560–572, and no. 132 (December 1840): 684–694.

Buchanan, Thomas C. *Black Life on the Mississippi: Slaves, Free Blacks, and the Western Steamboat World*. Chapel Hill: University of North Carolina Press, 2004.

Bulmuş, Birsen. *Plague, Quarantines, and Geopolitics in the Ottoman Empire*. Edinburgh: Edinburgh University Press, 2012.

Burgess Jr., Douglas R. *Engines of Empire: Steamships and the Victorian Imagination*. Stanford: Stanford University Press, 2016.

Burroughs, Robert. "Travel Writing and Rivers." In N. Das and T. Youngs (eds.), *The Cambridge History of Travel Writing*. Cambridge: Cambridge University Press, 2019, 330–344.

Buşe, Constantin. *Comerţul exterior prin Galaţi sub regim de port-franc (1837–1883)*. Bucureşti: Editura Academiei Republicii Socialiste România, 1976.

Butler, Rebecca. "'Can Any One Fancy Travellers without Murray's Universal Red Books?' Mariana Starke, John Murray and 1830s' Guidebook Culture." *Yearbook of English Studies* 48 (2018): 148–170.

Călători români paşoptişti. Edited by Dan Berindei. Bucureşti: Editura Sport-Turism, 1989.

Călători străini despre ţările române în secolul al XIX-lea. Serie nouă, vol. 2: *1822–1830*, edited by Paul Cernovodeanu and Daniela Buşă; vol. 3: *1831–1840*, edited by Daniela Buşă; vol. 4: *1841–1846*, edited by Daniela Buşă; vol. 5: *1847–1851*, edited by Daniela Buşă; vol. 6: *1852–1856*, editd by Daniela Buşă. Bucureşti: Editura Academiei Române, 2005–2010.

Căldăraru, Cristian-Dragoş. "Oraşul Galaţi în documentele din Manualul administrativ al Moldovei, 1834–1852." *Danubius* 32 (2014): 151–222.

Carlowitz, Aloïse-Christine de. "Voyage dans les Principautés Danubiennes." *Revue de Paris* 30 (September 15, 1856): 504–543.

Chahrour, Marcel. "'A Civilizing Mission'? Austrian Medicine and the Reform of Medical Structures in the Ottoman Empire, 1838–1850." *Studies in History and Philosophy of Science. Part C: Biological and Biomedical Sciences* 38, no. 4 (2007): 687–705.

Chircop, John, and Francisco Javier Martínez. "Introduction: Mediterranean Quarantine Disclosed: Space, Identity and Power." In John Chircop and Francisco Javier Martínez (eds.), *Mediterranean Quarantines, 1750–1914. Space, Identity and Power*. Manchester: Manchester University Press, 2018, 1–14.

Chrismar, F. S. *Skizzen eine Reise durch Ungarn in die Turkei*. Pest: Kilian Jun, 1834.

Cimpoeşu, Corina. "Călătoria la Constantinopol in anul 1857 a familiei Rosetti-Roznovanu." *Ion Neculce* 13–14 (2007–2009): 37–45.

Cioc, Mark. *The Rhine: An Eco-Biography, 1815–2000*. Seattle: University of Washington Press, 2002.

Cioroiu, Constantin. *Călători la Pontul Euxin*. Bucureşti: Editura Sport-Turism, 1984.

Claridge, R. T. *A Guide down the Danube*. London: F.C. Westley, 1839.

Claridge, R. T. *A Guide along the Danube from Vienna to Constantinople, Smyrna, Athens, the Morea, the Ionian Islands, and Venice, from the Notes of a Journey Made in the Year 1836*. London: F.C. Westley, 1837.

Codru-Drăguşanu, I. *Călătoriile unui român ardelean în ţară şi în străinătate (1835–44) ("Peregrinul transilvan")*. Edited by Constantin Onciu. Vălenii-de-Munte: Editura Tipografiei Societăţii Neamul Românesc, 1910.

Coons, Ronald E. *Steamships, Statesmen, and Bureaucrats: Austrian Policy towards the Steam Navigation Company of the Austrian Lloyd, 1836–1848.* Wiesbaden: Franz Steiner, 1975.

Corbin, Annalies. *The Material Culture of Steamboat Passengers: Archaeological Evidence from the Missouri River.* New York: Kluwer Academic, 2002.

Cornea, Paul. *Aproapele și departele.* București: Cartea Românească, 1990.

Cotoi, Călin. *Inventing the Social in Romania, 1848–1914. Networks and Laboratories of Knowledge.* Leiden: Brill, 2020. https://doi.org/10.30965/9783657704897.

Cumming, William Fullerton. *Notes of a Wanderer, in Search of Health, through Italy, Egypt, Greece, Turkey, up the Danube and down the Rhine.* London: Saunders and Otley; Edinburgh: Blackwood and Sons, 1839.

Dash, Comtesse. *Les vacances d'une Parisienne.* Paris: Calmann Lévy, 1890.

Delis, Apostolos. "Navigating Perilous Waters: Routes and Hazards of the Voyages to Black Sea in the Nineteenth Century." In Maria Christina Chatziioannou and Apostolos Delis (eds.), *Linkages of the Black Sea with the West: Navigation, Trade and Immigration.* Rethymno: Centre of Maritime History, Institute for Mediterranean Studies, Foundation of Research and Technology, 2020, 1–33.

Démidoff, Anatole de. *Voyage dans la Russie méridionale et la Crimée par la Hongrie, la Valachie et la Moldavie.* Paris: Ernest Bourdin, 1840.

Denkschrift der ersten k.k. privilegirten Donau-Dampfschiffahrts-Gesellschaft zur Erinnerung ihres fünfzigjährigen Bestandes. Wien: Selbstverlag der Gesellschaft, 1881.

Die Dampschiffahrt-Gesellschaft des Oesterreichisch-Ungarischen Lloyd von ihrem Entstehen bis auf unsere Tage 1836–1886. Trieste: Oesterreichisch-Ungarischen Lloyd, 1886.

Din relațiile și corespondența poetului Gheorghe Sion cu contemporanii săi. Edited by Ștefan Meteș. Cluj: Tipografia "Palla," 1939.

Din vremea renașterii naționale a Țării Românești: Boierii Golești, vol. 3: *1850–1852.* Edited by George Fotino. București: Imprimeria Națională, 1939.

Djuvara, Neagu. *Le pays roumain entre Orient et Occident: les principautés danubiennes au début du XIXe siècle.* Paris: Publications orientalistes de France, 1989.

Docea, Vasile. *Relații româno-germane timpurii: împliniri și eșecuri în prima jumătate a secolului XIX.* Cluj-Napoca: Presa Universitară Clujeană, 2000.

Dosch, Franz. *180 Jahre Donau-Dampfschiffahrts-Gesellschaft.* Erfurt: Sutton, 2009.

Drace-Francis, Alex. *The Making of Mămăligă: Transimperial Recipes for a Romanian National Dish.* Budapest: Central European University Press, 2022. https://doi.org /10.1515/9789633865842.

Drace-Francis, Alex. *The Traditions of Invention: Romanian Ethnic and Social Stereotypes in Historical Context.* Leiden: Brill, 2013.

Dusinberre, Martin, and Roland Wenzlhuemer. "Editorial: Being in Transit—Ships and Global Incompatibilities." *Journal of Global History* 11, no. 2 (2016): 155–162. https://doi. org/10.1017/S1740022816000036.

Edwards, Paul N. "Infrastructure and Modernity: Force, Time, and Social Organization in the History of Sociotechnical Systems." *Modernity and Technology* 1 (2003): 185–226.

Ekirch, A. Roger. *At Day's Close: Night in Times Past.* New York: W.W. Norton, 2006.

Elden, Stuart. "Plague, Panopticon, Police." *Surveillance & Society* 1, no. 3 (2003): 240–253.

Elias, Norbert. "Technization and Civilization." *Theory, Culture and Society* 12, no. 3 (1995): 7–42.

Elliott, Charles Boileau. *Travels in the Three Great Empires of Austria, Russia, and Turkey*, vol. 1. London: Bentley, 1838.

Emilciuc, Andrei. "The Trade of Galaţi and Brăila in the Reports of Russian Officials from Sulina Quarantine Station." In Constantin Ardeleanu and Andreas Lyberatos (eds.), *Port Cities of the Western Black Sea Coast and the Danube: Economic and Social Development in the Long Nineteenth Century*. Corfu: se, 2016, 63–93.

Ender, Thomas. *Die Wundermappe der Donau oder das Schönste und Merkwürdigste an den Ufern dieses Stromes vom Ursprung bis zur Mündung*. Pest: se, 1839.

Engelmann, Lukas, and Christos Lynteris. *Sulphuric Utopias: A History of Maritime Fumigation*. Cambridge, MA: MIT Press, 2020.

Epelde, Kathleen. *Travel Guidebooks to India: A Century and a Half of Orientalism*. PhD thesis. English Studies Program, University Wollongong, 2004. http://ro.uow.edu.au /theses/195 (accessed August 18, 2023).

Epure, Violeta-Anca. "Aspecte de viaţă cotidiană în principatele române prepaşoptiste surprinse de consulii şi voiajorii francezi: aşezările, casele, arhitectura." *Cercetări Istorice* 37 (2018): 271–288.

Epure, Violeta-Anca. "Imaginea femeii din Principatele Române în perioada prepaşoptistă în viziunea consulilor şi călătorilor francezi." *Terra Sebus. Acta Musei Sabesiensis* 5 (2013): 403–416.

Ersoy, Nermin, Yuksel Gungor, and Aslihan Akpinar. "International Sanitary Conferences from the Ottoman Perspective (1851–1938)." *Hygiea Internationalis* 10, no. 1 (2011): 53–79. https://doi.org/10.3384/hygiea.1403-8668.1110153.

Erste Donau-Dampfschiffahrts-Gesellschaft. *125 Jahre Erste Donaudampfschiffahrtsgesellschaft*. Wien: Donaudampfschiffahrtsgesellschaft, 1954.

Faifer, Florin. *Semnele lui Hermes. Memorialistica de călătorie (până la 1900) între real şi imaginar*. Bucureşti: Minerva, 1993.

Filimon, Nicolae. *Escursiuni în Germania meridională. Nuvele*. Edited by Paul Cornea. Bucureşti: Minerva, 1984.

Forsdick, Charles, Zoë Kinsley, and Kate Walchester (eds.). *Keywords for Travel Writing Studies. A Critical Glossary*. London: Anthem Press, 2019.

Foucault, Michel. *Discipline and Punish: The Birth of the Prison*. Translated from the French by Alan Sheridan. London: Penguin Books, 1977.

Foucault, Michel. "Of Other Spaces." *Diacritics* 16, no. 1 (1986): 22–27.

Frankland, Charles Colville. *Travels to and from Constantinople, in the Years 1827 and 1828*. London: Henry Colburn, 1829.

Fraser, James Baillie. *Narrative of the Residence of the Persian Princes in London, in 1835 and 1836, with an Account of Their Journey from Persia, and Subsequent Adventures*, vol. 2. London: Richard Bentley, 1838.

Friedberg-Mírohorský, Emanuel Salomon. *De la Praga la Focşani. Pe Dunăre spre România. Amintiri din sejurul militar în Principatul Valah din anul 1856*. Translated and edited by Anca Irina Ionescu. Bucureşti: Editura Lider, 2015.

Fynes, Clinton C. F. "A Few Pages from My Journal in Greece, Turkey, and the Danube." *Bentley's Miscellany* 16 (1844): 337–342.

Gane, Constantin. *Domniţa Alexandrina Ghica şi contele d'Antraigues*. Bucureşti: Editura Ziarului Universul, 1937.

Gatejel, Luminita. *Engineering the Lower Danube: Technology and International Cooperation in an Imperial Borderland*. Budapest: Central European University Press, 2022.

Gatejel, Luminita. "Overcoming the Iron Gates: Austrian Transport and River Regulation on the Lower Danube, 1830s–1840s." *Central European History* 49, no. 2 (2016): 162–180. https://doi.org/10.1017/S0008938916000327.

Gheorghe, Elisabeta. "Între etic și estetic. De ce călătoresc româncele? De ce scriu." *Studii și cercetări științifice. Seria filologie* 34 (2015): 39–53.

Gheorghiu, Emil. "Fumigația, ca mijloc de dezinfecție în carantinele din Țara Românească." In G. Brătescu (ed.), *Din istoria luptei antiepidemice în România. Studii și note.* București: Editura Medicală, 1972, 311–313.

Ghibănescu, Gh. (ed.). *Surete și izvoade (Documente slavo-române)*, vol. 10: *Documente cu privire la familia Râșcanu*. Iași: Tipografia Dacia, 1915.

Ghika, Aurélie. *La Valachie moderne*. Paris: Comon, 1850.

Girardin, Saint Marc. *Souvenirs de voyages et d'études*, vol. 1. Bruxelles: Delevingne et Callewaert, 1852.

Giuglea, G. "Călătoriile călugărului Chiriac dela Mănăstirea Secul. Călătoria la Muntele Atos și Ierusalim." *Biserica Ortodoxă Română* 54, nos. 3–4 and 11–12 (1936): 153–182, 697–720.

Gleason, J. H. *The Genesis of Russophobia in Great Britain 1815–1841*. Cambridge, MA: Harvard University Press, 1950.

Gleig, George Robert. *Germany, Bohemia, and Hungary Visited in 1837*, vol. 3. London: J.W. Parker, 1839.

Golescu, Constantin. *Însemnare călătoriei mele în anii 1824, 1825 și 1826*. Edited by Petre V. Haneș. București: Minerva, 1915.

Goodwin, Gráinne, and Gordon Johnston. "Guidebook Publishing in the Nineteenth Century: John Murray's Handbooks for Travellers." *Studies in Travel Writing* 17, no. 1 (2013): 43–61.

Green, Nicola. "On the Move: Technology, Mobility, and the Mediation of Social Time and Space." *Information Society* 18, no. 4 (2002): 281–292.

Gregory, Derek. *Geographical Imaginations*. Cambridge, MA: Blackwell, 1994.

Grössing, Helmuth, Johannes Binder, Ernst-Ulrich Funk, and Manfred Sauer. *Rot-Weiss-Rot auf Blauen Wellen*. Wien: Erste Donau-Dampfschiffahrts-Gesellschaft, 1979.

Guldi, Joanna. *Roads to Power: Britain Invents the Infrastructure State*. Cambridge, MA: Harvard University Press, 2012.

Hahn-Hahn, Ida von. *Letters from the Orient, or Travels in Turkey, the Holy Land and Egypt*. Translated from the German by S. Phillips. London: J.C. Moore, 1845.

Hajdu, Ada. "Dezvoltarea urbană a Băilor Herculane în secolul al XIX-lea și arhitectura stabilimentelor de băi." In *Spicilegium. Studii și articole în onoarea prof. Corina Popa*. București: Editura Unarte, 2015, 219–233.

Hajnal, Henry. *The Danube: Its Historical, Political and Economic Importance*. The Hague: Nijhoff, 1920.

A Handbook for Travellers in the Ionian Islands, Greece, Turkey, Asia Minor, and Constantinople. London: John Murray, 1840.

A Handbook for Travellers in Southern Germany. London: John Murray and Son, 1837.

A Handbook for Travellers in Southern Germany. Second edition, corrected and enlarged. London: John Murray, 1840.

A Handbook for Travellers in Southern Germany. Third edition, corrected and enlarged. London; John Murray, 1844.

A Handbook for Travellers in Southern Germany. Fifth edition. London: John Murray, 1850.

A Handbook for Travellers in Southern Germany. Sixth edition, corrected and enlarged. London: John Murray, 1853.

A Handbook for Travellers in Southern Germany. Seventh edition, corrected and enlarged. London: John Murray, 1857.

A Handbook for Travellers in Southern Germany. Eighth edition, corrected and enlarged. London: John Murray, 1858.

Harcourt, Freda. *Flagships of Imperialism: The P&O Company and the Politics of Empire from Its Origins to 1867*. Manchester: Manchester University Press, 2006.

Harrison, Mark. "Disease, Diplomacy and International Commerce: The Origins of International Sanitary Regulation in the Nineteenth Century." *Journal of Global History* 1, no. 2 (2006): 197–217. https://doi.org/10.1017/S1740022806000131.

Hart, Douglas. "Sociability and 'Separate Spheres' on the North Atlantic: The Interior Architecture of British Atlantic Liners, 1840-1930." *Journal of Social History* 44, no. 1 (2010): 189–212.

Hartleben, Conrad Adolph. *Panorama der oesterreichischen Monarchie, oder Malerisch-romantisches Denkbuch der schönsten und merkwürdigsten Gegenden derselben, der Gletscher, Hochgebirge, Alpenseen ... Schlösser, Burgen und Ruinen, so wie der interessantesten Donau-Ansichten*. Pest: Hartleben, 1839–1840.

Hartulari, Elena. "Istoria vieții mele de la anul 1801 (2)." *Convorbiri Literare* 10 (1926): 841–855.

Heinen, Armin. "Despre cultura tehnică a epocii moderne occidentale și perceperea cu totul diferită a timpului în România. Măsurarea timpului și timpul social din Evul Mediu până în prezent." *Analele Universității "Dunărea de Jos" Galați* 19, no. 7 (2008): 241–254.

Hillier, Bill. "The City as a Socio-technical System: A Spatial Reformulation in the Light of the Levels Problem and the Parallel Problem." In S. M. Arisona, G. Aschwanden, J. Halatsch, and P. Wonka (eds.), *Digital Urban Modeling and Simulation. Communications in Computer and Information Science*. Berlin: Springer, 2012, 24–48.

Holthaus, P. D. *Neue Reisen vollführt in den Jahren 1842–1845: Abenteuer und Beobachtungen*. Bremen: Langewiesche, 1846.

Huangfu, Day Jenny. "From Fire-Wheel Boats to Cities on the Sea: Changing Perceptions of the Steamships in the Late Qing, 1830s–1900s." *Australasian Journal of Victorian Studies* 20, no. 1 (2015): 50–63.

Hunter, Louis C. *Steamboats on the Western Rivers: An Economic and Technological History*. Newburyport: Dover, 2012.

Iacob, Dan Dumitru. "Călătoria lui Nicolae Rosetti-Roznovanu la Paris, în 1853." *Analele Științifice ale Universității "Alexandru Ioan Cuza" din Iași, s.n. Istorie* 63 (2017): 349–397.

Iacob, Dan Dumitru. "Divertisment și sociabilitate în principatele romane din prima jumătate a secolului al XIX-lea. Jocuri de societate." *Anuarul Institutului de Cercetări Socio-Umane Sibiu* 13–14 (2006–2007): 115–129.

Iacob, Dan Dumitru. "Premise pentru o istorie a formelor de sociabilitate mondenă. Salonul boieresc din prima jumătate a secolului XIX." *Xenopoliana* 10, nos. 1–4 (2002): 80–87.

Iancovici, L. "Un proiect pentru înlesnirea navigației pe Dunăre la Porțile de Fier din anul 1833." *Analele Universității București. Istorie* 23, no. 1 (1974): 189–191.

Ieromonahul, Andronic. *Călătoria la Muntele Athos (1858–1859)*. Edited by Petronel Zahariuc. Iași: Editura Universității "Alexandru Ioan Cuza" Iași, 2015.

Ionescu, Adrian-Silvan. "Experiența carantinei dunărene în notele de voiaj a doi călători americani: Vincent Otto Nolte și Valentine Mott." *Studii și Materiale de Istorie Modernă* 17 (2004): 57–68.

Iordachi, Constantin. *Liberalism, Constitutional Nationalism, and Minorities: The Making of Romanian Citizenship, c. 1750–1918*. Leiden: Brill, 2019.

Isar, Nicolae. *Sub semnul romantismului: de la domnitorul Gheorghe Bibescu la scriitorul Simeon Marcovici*. București: Editura Universității din București, 2003. http://ebooks .unibuc.ro/istorie/isar/index.htm (accessed August 28, 2023).

Jensen, John H., and Gerhard Rosegger. "British Railway Builders along the Lower Danube, 1856–1869." *Slavonic and East European Review* 46, no. 106 (1968): 105–128.

Jensen, John H., and Gerhard Rosegger. "Transferring Technology to a Peripheral Economy: The Case of the Lower Danube Transport Development, 1856–1928." *Technology and Culture* 19 (1978): 675–702.

Joanne, Adolphe. *Voyage en Orient*, vol. 1. Bruxelles: Delevingne et Callewaert imprimeurs-éditeurs, 1850.

Jokai, Maurus [Mór Jókai]. *Timar's Two Worlds (Az arany ember)*. Translated by [Agnes] Hegan Kennard. New York: D. Appleton, 1895.

Karović, Gordana. "Establishing Steam Navigation in the Principality of Serbia." In Dragana Amedoski (ed.), *Belgrade 1521–1867*. Belgrad: s.e., 2018, 383–406.

Kaufmann, Vincent. "On Transport History and Contemporary Social Theory." *Journal of Transport History* 28, no. 2 (2007): 302–306.

King, Charles. *The Black Sea: A History*. Oxford: Oxford University Press, 2004.

Király, Edit. *"Die Donau ist die Form": Strom-Diskurse in Texten und Bildern des 19. Jahrhunderts*. Wien: Böhlau, 2017. https://doi.org/10.26530/oapen_629710.

Koch, Karle Heinrich Emil. *Wanderungen im Oriente, während der Jahre 1843 und 1844: Reise längs der Donau nach Konstantinopel und nach Trebisond*, vol. 1. Weimar: Druck und Verlag des Landes-Industrie-Comptoirs, 1846.

Kogălniceanu, M. *Scrisori, 1834–1849*. Edited by Petre V. Haneș. București: Minerva, 1913.

Kogălniceanu, Mihail. *Opere*, vol. 1: *Beletristica, studii literare, culturale si sociale*. Edited by Dan Simonescu. București: Editura Academiei Republicii Socialiste România, 1974.

Kohl, Johann Georg. *Austria: Vienna, Prague, Hungary, Bohemia, and the Danube; Galicia, Styria, Moravia, Bukovina, and the Military Frontier*. London: Chapman and Hall, 1843.

Kohn, Rita (ed.). *Full Steam Ahead: Reflections on the Impact of the First Steamboat on the Ohio River, 1811–2011*. Indianapolis: Indiana Historical Society Press, 2011.

Konachi, Costachi. *Poesii: alcătuiri și tălmăciri*. Ediția a doua. Iași: Editura Librăriei Frații Saraga, 1887.

Koshar, Rudy. "'What Ought to Be Seen': Tourists' Guidebooks and National Identities in Modern Germany and Europe." *Journal of Contemporary History* 33, no. 3 (1998): 323–340.

Kostova, Ludmila. "Meals in Foreign Parts: Food in Writing by Nineteenth-Century British Travellers to the Balkans." *Journeys* 4, no. 1 (2003): 21–44.

Kunike, Adolph (ed.). *Zwey hundert vier und sechzig Donau-Ansischen nach dem Laufe des Donaustroms, von seinem Ursprunge bis zu seinem Ausflusse in das schwarze Meer. Sammt einer Donaukarte*. Wien: Auf Kosten des Herausgebers, gedruckt bey L. Grund, 1826.

Kunisch, Richard. *Bukarest und Stambul. Skizzen aus Ungarn, Rumunien [sic] und der Türkei*. Berlin: Nicolaische Verlagsbuchhandlung, 1861.

Labatut, Auguste. "Bucharest et Jassy." *Revue de Paris* 50 (1838): 143–158.

Lăcusteanu, Grigore. *Amintirile colonelului Lăcusteanu.* Edited by Rodica Pandele Peligrad. Iași: Polirom, 2015.

Laderman, Scott. "Guidebooks." In Carl Thomas (ed.), *The Routledge Companion to Travel Writing.* London: Routledge, 2015, 278–288.

Lavalléé, Théophile-Sébastien. "Les villes du Bas-Danube." *Revue d'Orient* 14 (1853): 400–407.

Lefebvre, Thibault. *Études diplomatiques et économiques sur la Valachie.* Paris: Guillaumin et Cié, 1858.

Mackintosh, Will. "'Ticketed Through': The Commodification of Travel in the Nineteenth Century." *Journal of the Early Republic* 32, no. 1 (2012): 61–89. https://doi.org/10.1353/jer.2012.0001.

Manta-Cosma, Ioana. "O cafea, o lulea și o iubită. O iubire 'pașoptistă' în Veneția anului 1846." *Caiete de Antropologie Istorică* 19 (2011): 15–29.

Marchebeus, [Jean-Baptiste]. *Voyage de Paris à Constantinople par bateau à vapeur.* Paris: Arthus Bertrand, 1839.

Marmier, Xavier. *Du Rhin au Nil: Tyrol–Hongrie–Provinces Danubiennes–Syrie–Palestine–Egypte. Souvenirs de voyages,* vol. 1. Bruxelles: Delevingne et Callewaert, 1852.

Mărturii istorice privitoare la viața și domnia lui Știrbei Vodă. Edited by Nicolae Iorga. București: Institutul de Arte Grafice și Editură Minerva, 1905.

Mârza, Radu. *Călători români privind pe fereastra trenului. O încercare de istorie culturală.* Iași: Polirom, 2020.

Mason, John. *Three Years in Turkey: The Journal of a Medical Mission to the Jews.* London: John Snow, 1860.

Memoriile principelui Nicolae Suțu: mare logofăt al Moldovei, 1798–1871. Edited by Georgeta Penelea Filitti. București: Editura Fundației Culturale Române, 1997.

Mevissen, Robert. *Constructing the Danube Monarchy: Habsburg State-Building in the Nineteenth Century.* PhD thesis. Georgetown University, 2017.

Mevissen, Robert Shields. "Meandering Circumstances, Fluid Associations: Shaping Riverine Transformations in the Late Habsburg Monarchy." *Austrian History Yearbook* 49 (2018): 23–40. https://doi.org/10.1017/S0067237818000061.

Meyer, Joseph. *Donau-Ansichten von der Quelle des Stroms bis zu seiner Mündung: nach Original-Zeichnungen,* Hildburghausen: Bibliogr. Inst., 1838–1839.

Minescu, Constantin N. *Istoria poștelor române. Originea, desvoltarea și legislațiunea lor.* București: Imprimeria Statului, 1916.

Mislin, Jacques. *Saint Lieux. Pèlerinage à Jérusalem en passant par l'Autriche, la Hongrie, les Provinces Danubiennes, Constantinople, l'Archipel, le Liban, la Syrie, Alexandrie, Malte, la Sicile et Marseille.* Par Mgr. Mislin, prélat de la Maison de Sa Sainteté Pie IX, vol. 1. Paris: Guyot, 1851.

Mitchell, W. J. Thomas. *Landscape and Power.* Chicago: University of Chicago Press, 2002.

Moltke, Helmuth von. "Tagebuch der Reise nach Konstantinopel." In *Gesammelte Schriften und Denkwürdigkeiten des General-Feldmarschalls Grafen Helmuth von Moltke,* vol. 1. Berlin: Ernst Siegfried Mittler, 1892, 103–139.

Monson, Frederick John. *Journal of a Tour in Germany.* London: Printed by W.H. Dalton, 1840.

Moraglio, Massimo. "Seeking a (New) Ontology for Transport History." *Journal of Transport History* 38, no. 1 (2017): 3–10. https://doi.org/10.1177/0022526617709168.

Morot, Jean-Baptiste. *Journal de voyage. Paris à Jérusalem. 1839 et 1840.* Deuxième édition. Paris: Impr. J. Claye, 1873.

Mott, Valentine. *Travels in Europe and the East. Embracing Observations Made during a Tour through Great Britain, Ireland, France, Belgium, Holland, Prussia, Saxony, Bohemia, Austria, Bavaria, Switzerland, Lombardy, Tuscany, the Papal States, the Neapolitan Dominions, Malta, the Islands of the Archipelago, Greece, Egypt, Asia Minor, Turkey, Moldavia, Wallachia, and Hungary in the Years 1834, '35, '37, '38, '39, '40 and '41.* New York: Harper & Brothers, 1842.

Müller, Adalbert. *Die Donau vom Ursprunge bis zu den Mündungen,* vol. 2: *Die untere Donau.* Regensburg: Manz, 1841.

Munteanu, Alina Cristina. "Travelling in Oriental Romania in the Second Half of the Nineteenth Century, According to the Writings of Western Travellers." *Studia Universitatis Petru Maior. Historia* 15, no. 1 (2015): 15–24.

Murgescu, Bogdan. *Țările Române între Imperiul Otoman și Europa creștină.* Iași: Polirom, 2012.

Nardi, Francesco. *Ricordi di un viaggio in Oriente.* Roma: Stamp. della S.C. de Propaganda Fide, 1866.

Năstase, Gheorghe, Cristina Ionescu, and Rodica Anghel. "Câteva informații despre carantina de la Galați." In G. Brătescu (ed.), *Din istoria luptei antiepidemice în România. Studii și note.* București: Editura Medicală, 1972, 309–310.

Negru, Ion. "Cum vedea doctorul Pavel Vasici carantinele în 1847." In G. Brătescu (ed.), *Din istoria luptei antiepidemice în România. Studii și note.* București: Editura Medicală, 1972, 315–322.

Negulescu, Paul, and George Alexianu. *Regulamentele Organice ale Valahiei și Moldovei.* București: Tipografia Eminescu, 1944.

Nicolson, Miklos Szucs. "Count Istvan Széchenyi (1792–1860): His Role in the Economic Development of the Danube Basin." *Explorations in Economic History* 6, no. 3 (1954): 163–180.

Noyes, James O. *Roumania: Border Land of the Christian and the Turk, Comprising Adventures of Travel in Eastern Europe and Western Asia.* New York: Rudd & Carleton, 1857.

O'Brien, Patrick. *Journal of a Residence in the Danubian Principalities in the Autumn and Winter of 1853.* London: Richard Bentley, 1854.

Offermann, Michael David, and Roland Wenzlhuemer. "Ship Newspapers and Passenger Life aboard Transoceanic Steamships in the Late Nineteenth Century." *Transcultural Studies* 3, no. 1 (2012): 77–121.

Olin, Stephen. *Travels in Egypt, Arabia Petræa, and the Holy Land,* vol. 2. New York: Harper and Brothers, 1843.

Oliphant, Laurence. *The Russian Shores of the Black Sea in the Autumn of 1852: With a Voyage down the Volga, and a Tour through the Country of the Don Cossacks.* Edinburgh: William Blackwood and Sons, 1853.

Osterhammel, Jürgen. *The Transformation of the World: A Global History of the Nineteenth Century.* Translated by Patrick Camiller. Princeton: Princeton University Press, 2015.

Oțetea, Andrei. *Pătrunderea comerțului românesc în circuitul internațional (în perioada de trecere de la feudalism la capitalism).* București: Editura Academiei Republicii Socialiste România, 1977.

The Overland Mail and the Austrian Lloyd's. London: G. Mann and F. Thimm, 1847.

Pajură, C., and D. T. Giurescu. *Istoricul orașului Turnu-Severin (1833–1933).* București: Tiparul Românesc, 1933.

Păltănea, Paul. *Istoria orașului Galați de la origini până la 1918*, vol. 1. Second edition. Edited by Eugen-Dan Drăgoi. Galați: Editura Partener, 2008.

Păltănea, Paul. *Viața lui Costache Negri*. Iași: Junimea, 1985.

Panzac, Daniel. *La Peste dans l'Empire ottoman 1700–1850*. Leuven: Ed. Peeters, 1985.

Panzac, Daniel. "Politique sanitaire et fixation des frontières: l'exemple Ottoman (XVIII-XIX siècles)." *Turcica* 31 (1999): 87–108.

Papacostea, Șerban. "Drum și stat." *Studii și materiale de istorie medie* 10 (1983): 9–55.

Papastate, Constantin D. *Vasile Alecsandri și Elena Negri: cu un Jurnal inedit al poetului*. București: Tiparul Românesc, 1947.

Pardoe, Julia. *The City of the Sultan: And Domestic Manners of the Turks in 1836*, vol. 2. London: Henry Colburn, 1837.

Pascu, Ștefan, Radu Pantazi, and Teofil Gridan. *Istoria gîndirii și creației științifice și tehnice românești*. București: Editura Academiei Republicii Socialiste România, 1982.

Paskaleva, Virginia. "Le rôle de la navigation à vapeur sur le bas Danube dans rétablissement de liens entre l'Europe Centrale et Constantinople jusqu'à la guerre de Crimée." *Bulgarian Historical Review* 4, no, 1 (1976): 64–76.

Paskaleva, Virginia. "Shipping and Trade on the Lower Danube in the Eighteenth and Nineteenth Centuries." In Apostolos E. Vacalopoulos, Constantinos D. Svolopoulos, and Béla K. Király (eds.), *Southeast European Maritime Commerce and Naval Policies from the Mid-Eighteenth Century to 1914*. Boulder: Social Science Monographs and Atlantic Research and Publications, 1988, 131–151.

Păun-Constantinescu, Ilinca. *Shrinking Cities in Romania. 1. Research and Analysis = Orașe românești în declin. 1. O cercetare critică*. Berlin: DOM Publishers and MNAC, 2019.

Pelimon, Alexandru. *Impresiuni de călătorie în România*. Edited by Dalila-Lucia Aramă. București: Editura Sport-Turism, 1984.

Penelea, Georgeta. "Organizarea carantinelor în epoca regulamentară." *Studia Universitatis Babeș-Bolyai* 14 (1969): 29–41.

Perthes, Jacques Boucher de. *Voyage à Constantinople par l'Italie, la Sicilie et la Grèce. Retour par la Mer Noire, la Roumelie, la Bulgarie ... en Mai, Juin, Juillet et Aout 1853*, vol. 2. Paris: Treuttel et Wurtz, 1855.

Petrescu, Ghenadie, Dimitrie A. Sturdza, and Dimitrie C. Sturdza (eds.). *Acte și documente relative la istoria renascerei României*, vol. 3. Bucuresci: Göbl, 1889.

Petrescu, Ștefan. "Migrație și carantine în porturile dunărene: controlul documentelor de călătorie în epoca Regulamentelor Organice." *Studii și materiale de istorie modernă* 25 (2012): 97–116.

Petrică, Virginia. *Topography of Taste: Landmarks of Culinary Identity in the Romanian Principalities from the Perspective of Foreign Travellers*. București: Editura Academiei Române, 2018.

Pfeiffer, Ida. *Visit to the Holy Land, Egypt, and Italy*. Translated from the German by H. W. Dulcken. London: Ingram Cooke, 1852.

Pigeory, Félix. *Les pèlerins d'Orient, Lettres artistiques et historiques sur un voyage dans les Provinces Danubiennes, la Turquie, la Syrie et la Palestine*. Paris: E. Dentu, 1854.

Ploner, Josef. "Tourist Literature and the Ideological Grammar of Landscape in the Austrian Danube Valley, ca. 1870–1945." *Journal of Tourism History* 4, no. 3 (2012): 237–257. https://doi.org/10.1080/1755182X.2012.711376.

Popa, Bogdan. "Experiența fizică a frontierei: carantina." In Romanița Constantinescu (ed.), *Identitate de frontieră în Europa lărgită: Perspective comparate*. Iași: Polirom, 2008, 93–101.

Popa, Marian. *Călătoriile epocii romantice*. București: Editura Univers, 1972.

Popa, Mircea. "George Bariț—călătorul." *Anuarul Institutului de Istorie "George Barițiu" din Cluj-Napoca. Series Historica* 42 (2003): 89–99.

Popova-Nowak, Irina V. "The Odyssey of National Discovery: Hungarians in Hungary and Abroad, 1750–1850." In Wendy Bracewell and Alex Drace-Francis (eds.), *Under Eastern Eyes: A Comparative Introduction to East European Travel Writing in Europe*. Budapest: Central European University Press, 2008, 195–222.

Popovici, V. C., C. Anghelescu, and L. Boicu. *Dezvoltarea economiei Moldovei între anii 1848 și 1864: contribuții*. București: Editura Academiei Republicii Populare Romîne, 1963.

Potra, George. *Călători români în țări străine*. București: Tip. Oltenia, 1939.

Potra, George. "Călătoria unui boier moldovean în Europa la mijlocul secolului al XIX-lea." *Revista istorică* 19, nos. 4–6 (1933): 126–139.

Potra, George. "Statele Europei la 1846–1847, văzute de un boier moldovean." *Revista istorică română* 9 (1939): 207–245.

Pratt, Mary Louise. "Arts of the Contact Zone." *Profession* (1991): 33–40.

Promitzer, Christian. "Prevention and Stigma: The Sanitary Control of Muslim Pilgrims from the Balkans, 1830–1914." In John Chircop and Francisco Javier Martínez (eds.), *Mediterranean Quarantines, 1750–1914: Space, Identity and Power*. Manchester: Manchester University Press, 2018, 145–169.

Pungă, Doina. "Litografia europeană în secolul al XIX-lea. Repere istorice, tehnice și stilistice, funcționalitate." *Muzeul Național* 15 (2003): 120–140.

Quin, Michael. *A Steam Voyage Down the Danube: With Sketches of Hungary, Wallachia, Servia, and Turkey etc*, vol. 1. Second edition, revised and corrected. London: Richard Bentley, 1835.

Rădvan, Laurențiu (ed.). "Drumuri de țară și drumuri de oraș în Țara Românească în secolele XVII–XVIII." In *Orașul din spațiul românesc între Orient și Occident*. Iași: Editura Universității "Alexandru Ioan Cuza," 2007, 67–114.

Ralet, Dimitrie. *Suvenire și impresii de călătorie în România, Bulgaria, Constantinopole*. Edited by Mircea Anghelescu. București: Minerva, 1979.

Râpă-Buicliu, Dan, and Iulian Capsali. "Însemnări din 'Jurnalul de călătorie în Occident' al boierului moldovean Iancu Prăjescu." *Danubius* 27 (2009): 235–262.

Reichard, Heinrich August Ottokar. *An Itinerary of Germany … To Which Is Added an Itinerary of Hungary and Turkey*. Paris: A. and W. Galignani, 1826.

Rennella, Mark, and Whitney Walton. "Planned Serendipity: American Travelers and the Transatlantic Voyage in the Nineteenth and Twentieth Centuries." *Journal of Social History* 38, no. 2 (2004): 365–383.

Rey, William. *Autriche, Hongrie et Turquie. 1839–1848*. Paris: Joel Cherbuliez, 1849.

Ricketts, Clemuel Green. *Notes of Travel: In Europe, Egypt, and the Holy Land, Including a Visit to the City of Constantinople, in 1841 and 1842*. Philadelphia: C. Shermann, 1844.

Riedler, Florian, and Nenad Stefanov (eds.). *The Balkan Route: Historical Transformations from Via Militaris to Autoput*. Berlin: Walter de Gruyter, 2021.

Robarts, Andrew. *A Plague on Both Houses? Population Movements and the Spread of Disease across the Ottoman-Russian Black Sea Frontier, 1768–1830s*. PhD thesis. Georgetown University, 2010.

Romer, Mrs [Isabella Frances]. *The Bird of Passage or, Flying Glimpses of Many Lands*, vol. 2. London: R. Bentley, 1849.

Rosetti, C. A. "Note intime scrise zilnic (1844–1859)." In *Lui C.A. Rosetti: la o sută de ani dela naşterea sa*. Bucureşti: "Democraţia," Revista Cercului de Studii P.N.L., 1916.

Rosetti, Radu. *Amintiri. Ce am auzit de la alţii. Din copilărie. Din prima tinereţe*. Bucureşti: Humanitas, 2017.

Rothenberg, Gunther E. "The Austrian Sanitary Cordon and the Control of the Bubonic Plague: 1710–1871." *Journal of the History of Medicine and Allied Sciences* 28, no. 1 (1973): 15–23.

Said, Edward W. *Orientalism*. London: Routledge & Kegan Paul, 1978.

Sauer, Manfred. "Österreich und die Sulina-Frage, 1829–1854." 1 and 2. *Mitteilungen des Österreichischen Staatsarchivs* 40 (1987): 185–236, and 41 (1990): 72–137.

Schiffer, Reinhold. *Oriental Panorama: British Travellers in 19th Century Turkey*. Amsterdam: Rodopi, 1999.

Schivelbusch, Wolfgang. *The Railway Journey: The Industrialization of Time and Space in the Nineteenth Century*. Translated from the German. Oakland: University of California Press, 2014.

Schulz-Forberg, Hagen. "The Sorcerer's Apprentice: English Travellers and the Rhine in the Long Nineteenth Century." *Journeys* 3, no. 2 (2002): 86–110.

Scrisori vechi de studenţi (1822–1889). Edited by Nicolae Iorga. Bucureşti: s.e., 1934.

Šedivý, Miroslav. "From Hostility to Cooperation? Austria, Russia and the Danubian Principalities 1829–40." *Slavonic and East European Review* 89, no. 4 (2011): 630–661.

Sheller, Mimi, and John Urry. "The City and the Car." *International Journal of Urban and Regional Research* 24, no. 44 (2000): 737–757.

Sheller, Mimi, and John Urry. "The New Mobilities Paradigm." *Environment and Planning A* 38, no. 2 (2006): 207–226.

Sion, Gh. *Suvenire contimporane*. Bucureşti: Minerva, 1915.

Skene, Felicia. *Wayfaring Sketches among the Greeks and the Turks, and on the Shore of the Danube, by a Seven Years' Resident in Greece*. London: Chapman and Hall, 1847.

Skene, James Henry. *The Frontier Lands of the Christian and the Turk*, vol. 1. London: Richard Bentley, 1853.

Slade, Adolphus. *Travels in Germany and Russia, Including a Steam Voyage by the Danube and the Euxine from Vienna to Constantinople, in 1838–39*. London: Longman, Orme, Brown, Green, and Longmans, 1840.

Slăvescu, Victor (ed.). *Corespondenţa între Ion Ionescu dela Brad şi Ion Ghica 1846–1874*. Bucureşti: Imprimeria Naţională, 1943.

Smancini, Giorgio. *Scorsa piacevole in Grecia, Egitto, Turchia sul Danubio e da Vienna alla Lombardia*. Milano: Tipografia Manini, 1844.

Smyth, Warington Wilkinson. *A Year with the Turks or Sketches of Travel in the European and Asiatic Dominions of the Sultan*. New York: Redfield, 1854.

Snow, Robert. *Journal of a Steam Voyage down the Danube to Constantinople, and Thence by Way of Malta and Marseilles to London*. London: Moyes and Barclay, 1842.

Spencer, Edmund. *Travels in Circassia, Krim Tartary etc. Including a Steam Voyage down the Danube, from Vienna to Constantinople and Round the Black Sea in 1836*, vol. 1. London: Henry Colburn, 1837.

Stafford, Jonathan. "A Sea View: Perceptions of Maritime Space and Landscape in Accounts of Nineteenth-Century Colonial Steamship Travel." *Journal of Historical Geography* 55 (2017): 69–81. https://doi.org/10.1016/j.jhg.2016.09.006.

Stan, Constantin I., and Alexandru Gaiță. *Călătorii ale românilor în centrul și vestul Europei (1800–1848)*. Buzău: Editgraph, 2013.

Stăncescu, Scarlat A. *Din trecutul orașului Giurgiu*. București: Unirea, 1935.

Stelling-Michaud, Suzanne (ed.). *Le Livre du recteur de l'Académie de Genève: 1559–1878*, vol. I.V: *Notices biographiques des étudiants, N–S*. Genève: Librairie Droz, 1976.

Suțu, Nicolae. *Amintiri de călătorie, 1839–1847*. Edited by Petruța Spânu, Gheorghe Macarie, and Dumitru Scorțanu. Iași: Fides, 2001.

Tappe, E. D. "Was Quin's 'Moldavian Adventurer' Slugerul Burada?" *Revue des Études Sud-Est Européennes* 12, no. 4 (1974): 588–590.

Tinku-Szathmáry, Balázs. "Gőzhajóval a Dunán Bécsből Konstantinápolyig I." *Közlekedés-és technikatörténeti Szemle* (2018): 11–38.

Tinku-Szathmáry, Balázs. "Gőzhajóval a Dunán Bécsből Konstantinápolyig II." *Közlekedés-és technikatörténeti Szemle* (2019): 9–44.

Tipei, Alex R. "Audience Matters: 'Civilization-Speak', Educational Discourses, and Balkan Nationalism, 1800–1840." *European History Quarterly* 48, no. 4 (2018): 658–685. https://doi.org/10.1177/0265691418799547.

Todorova, Maria. *Imagining the Balkans*. Oxford: Oxford University Press, 2009.

Trăușan-Matu, Lidia. "Doctorul Nicolae Gussi și istoria carantinei în Țara Românească." *Jurnal Medical Brașovean* 1 (2020): 133–141.

Trăușan-Matu, Lidia, and Octavian Buda. "Cholera, Quarantines and Social Modernisation at the Danube Border of the Ottoman Empire: The Romanian Experience between 1830 and 1859." *Social History of Medicine* 36, no. 1 (2023): 24–41. https://doi.org/10.1093/shm/hkac064.

Urquhart, David. *The Mystery of the Danube*. London: Bradbury and Evans, 1851.

Urquhart, David. *Turkey and Its Resources*. London: Saunders and Otley, 1833.

Urry, John. "Social Networks, Mobile Lives and Social Inequalities." *Journal of Transport Geography* 21 (2012): 24–30. https://doi.org/10.1016/j.jtrangeo.2011.10.003.

Vaillant, Jean Alexandre. *La Roumanie ou histoire, langue, littérature, orographie, statistique des peuples de la langues d'ore, ardialiens, vallaques et moldaves, résumé sous le nom de Romains*, vol. 3. Paris: Arthus Bertrand, 1844.

Valon, Alexis de. *Une année dans le Levant*, vol. 2: *La Turquie sous Abdul-Medjid*. Paris: Jules Labitte, 1846.

Vane, Charles William, Marques of Londonderry. *A Steam Voyage to Constantinople by the Danube and Rhine in 1840–41 and to Portugal, Spain &c in 1839*, vol. 1. London: Henry Colburn, 1842.

Vane, Frances Anne. *Narrative of a Visit to the Courts of Vienna, Constantinople, Athens, Naples etc. by the Marchioness of Londonderry*. London: Henry Colburn, 1844.

Verne, Jules. *A Floating City*. Translated from the French. London: Argo, 1918.

Vezenkov, Alexander. "Entangled Geographies of the Balkans: The Boundaries of the Region and the Limits of the Discipline." In Roumen Daskalov, Diana Mishkova, Tchavdar Marinov, and Alexander Vezenkov (eds.), *Entangled Histories of the Balkans*, vol. 4: *Concepts, Approaches, and (Self-)Representations*. Leiden: Brill, 2017, 115–256.

Vintilă, Constanța. *Changing Subjects, Moving Objects: Status, Mobility, and Social Transformation in Southeastern Europe, 1700–1850*. Translated by James Christian Brown. Paderborn: Brill Schöningh, 2022.

Vintilă-Ghițulescu, Constanța. *Patimă și desfătare: despre lucrurile mărunte ale vieții cotidiene în societatea românească: 1750–1860*. București: Humanitas, 2015.

Vintilă-Ghiţulescu, Constanţa. *Tinereţile unui ciocoiaş. Viaţa lui Dimitrie Foti Merişescu de la Colentina scrisă de el însuşi la 1817.* Bucureşti: Humanitas, 2019.

Weedon, Alexis. "Blewitt, (John) Octavian (1810–1884), Writer and Literary Administrator." In *Oxford Dictionary of National Biography.* Oxford: Oxford University Press, 2004. https://doi.org/10.1093/ref:odnb/2645.

Wenzlhuemer, Roland. "The Ship, the Media, and the World: Conceptualizing Connections in Global History." *Journal of Global History* 11, no. 2 (2016): 163–186. https://doi.org/10.1017/S1740022816000048.

Wilkinson, William. *An Account of the Principalities of Wallachia and Moldavia with Various Political Observations Relating to Them.* London: Longman, Hurst, Rees, Orme, and Brown, 1820.

Williams, David M., and John Armstrong. "'One of the Noblest Inventions of the Age': British Steamboat Numbers, Diffusion, Services and Public Reception, 1812–c.1823." *Journal of Transport History* 35, no. 1 (2014): 18–34.

Williams, Simon. *The Politics of Sleep: Rights, Risks, and Regulations.* New York: Palgrave Macmillan, 2011.

Wolff, Larry. *Inventing Eastern Europe: The Map of Civilization on the Mind of the Enlightenment.* Stanford: Stanford University Press, 2004.

Zahariuc, Petronel. "Bacşişuri, mătăsuri şi argintării. Călătoria boierilor moldoveni la Constantinopol în 1822." In Dan Dumitru Iacob (ed.), *Avere, prestigiu şi cultură materială în surse patrimoniale. Inventare de averi din secolele XVI–XIX.* Iaşi: Editura Universităţii "Alexandru Ioan Cuza," 2015, 311–369.

Zahariuc, Petronel. "Călătoria ieromonahului Chiriac din Mănăstirea Secu la Muntele Athos (1840-1841)." *Analele Ştiinţifice ale Universităţii "Alexandru Ioan Cuza" din Iaşi, s.n. Istorie* 61 (2015): 249–264.

Zahariuc, Petronel. "Despre acelaşi ieromonah Andronic de la Mănăstirile Neamţ şi Secu, însă despre o altă călătorie: la Ierusalim (1859)." In Liliana Rotaru (ed.), *Historia est magistra vitae. Civilizaţie, valori, paradigme, personalităţi. In honorem profesor Ion Eremia.* Chişinău: Biblioteca Ştiinţifică Centrală, 2019, 234–243.

Zahariuc, Petronel (ed.). "Din corespondenţa unui călugăr român la Muntele Athos în secolul al XIX-lea: Maxim hagiul." In *Relaţiile românilor cu Muntele Athos şi cu alte Locuri Sfinte (sec. XIV–XX). In honorem Florin Marinescu.* Iaşi: Editura Universităţii "Alexandru Ioan Cuza" Iaşi, 2017, 183–256.

Zahariuc, Petronel. "Sur le hiéromoine Andronic des monastères de Neamţ et de Secu et sur son voyage au Mont Athos (1858–1859)." *Analele Ştiinţifice ale Universităţii "Alexandru Ioan Cuza" din Iaşi, s.n. Istorie* 62 (2016): 151–197.

Index

9 789633 867532